Supported by:

Cumbria Boglife, Natural England, the British Ecological Society Peatlands SIG,

UKECONET / Biodiversity Research Group (BRG), Cumbria Wildlife Trust,

Solway Wetlands Partnership, Solway Coast AONB,

Sheffield Hallam University & partners

PEATLANDS RESEARCH GROUP

British Ecological Society
Peatlands Research

History & Heritage of the Bogs & Peatlands of Cumbria

& Surrounding Areas

Edited by Ian D. Rotherham and Christine Handley

2021
Published by Wildtrack Publishing, Sheffield

ISSN 1354-0262

ISBN 978-1-904098-71-3

9 781904 098713

Venture House, 103 Arundel Street, Sheffield, S1 2TN

Landscape Archaeology and Ecology,
14

Contents

Industrial peat-cutting machinery © Ian Rotherham

History & Heritage of the Bogs & Peatlands of Cumbria & Surrounding Areas

Preamble

In November 2017, a two-day seminar was held at Burgh-by-Sands Parish Hall with Cumbria Boglife, Natural England, the British Ecological Society Peatlands SIG, UKECONET / Biodiversity Research Group (BRG), Cumbria Wildlife Trust, the RSPB, the Solway Wetlands Landscape Partnership, Solway Connections Guided Heritage Tours, & partners. There were local site visits organised by the local and regional partners.

The event explored the history and cultural aspects of peat bogs in and around Cumbria and the surrounding areas. Local people made a major contribution to the success of the event, presented at the indoor sessions, and provided displays and artefacts.

Themes and topics included:

1. The history of peat bogs and peatlands
2. The history of peat harvesting and usage
3. The oral history and memories
4. The heritage of peat cutting – tools, equipment, buildings, *etc*
5. The historic records
6. The archaeology of peat-cut sites
7. The conservation of peat cutting heritage
8. And more!

Peat moss litter 1960s © Ian Rotherham

Chapter 1. History and heritage in the bog – examples from Cumbria and the surrounding areas

Ian D. Rotherham
Sheffield Hallam University

Summary

This chapter introduces and summarises the occurrence and use of peat in and around Cumbria. It sets the scene in a wider social and economic context with examples from elsewhere. The account then describes with examples, some aspects of commercial or industrial peat exploitation and then of domestic or community usage.

Keywords: *peatlands, peat bogs, peat usage, peat history, peat heritage, peat fuel*

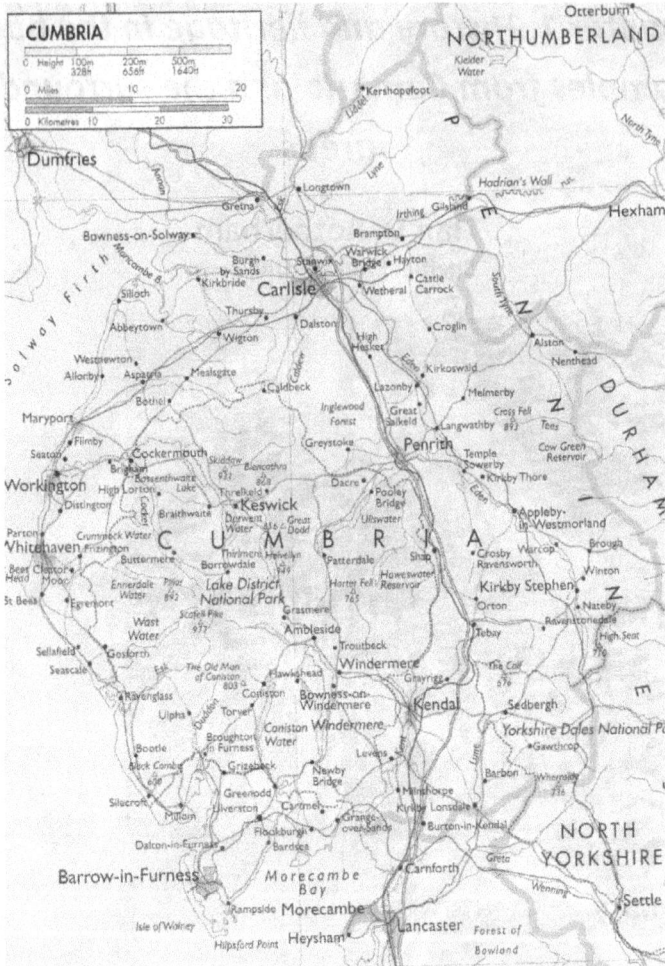

Figure 1. Map showing the study region

Introduction

Cumbria has a rich (perhaps uniquely so) literature concerning its countryside, its people, its heritage, and its natural resources. This documentation provides a fertile source of information on the landscape and its utilisation over the centuries. However, the detailed roles of peat bogs and peat have been generally neglected and, if not overlooked, at least not celebrated. William Rollinson (1981, 1996) notes peat cutting and provides some images of the practice in Cumbria, and Susan Denyer (1991) presents a vivid insight into peat barns or 'peat-scales' in the region. Carnie (2002) noted how domestic peat use was carefully managed with maps drawn up and regularly updated to show the allocations of peat-grounds to each household in a settlement such as for example, Snake Moss at Scandale. As each moss became exhausted it was important that another one be opened up. However, in this case with increasing population pressure the supply was virtually exhausted by the late 1700s.

Historically, peat landscapes in and around Cumbria have been very significant and their scale was sometimes immense. What we see today is a sadly shrunken resource. This abundance of peat is reflected in some of the histories such as the 1771 -1772 bog burst at Solway Moss described by the Welsh travel-writer Thomas Pennant, when the site of around 600 hectares extent discharged around 120 hectares in a violent eruption of wet peat. This was described as a 'Stygian tide' (i.e. like the River Styx - extremely dark, gloomy, or forbidding) of semi-fluid black peat up to four metres deep which engulfed several farmhouses though fortunately without loss of life. As was the case in other examples, the problem was triggered by both the extent of the peat growth at the time and its de-stabilisation by localised peat-cutting. Heavy rain building up to the burst was another contributing factor.

Another but different historic incident was the tragic Battle of Solway Moss in November 1542 when 15,000 to 18,000 Scottish troops were routed by 3,000 English. Following from Henry VIII's break from the Church in Rome, he had made incursions into Scotland which triggered a response from the Scots moving an army led by Lord Maxwell down the west coast. The Scots advance into England was met near Solway Moss by Lord Wharton and his 3,000 men. The battle was uncoordinated and was a rout. Sir Thomas Wharton described the battle as the overthrow of the Scots between the rivers Esk and Lyne. In the conflict the Scots were trapped south of the River Esk on English territory between the river and the Moss and following intense fighting surrendered themselves and their field guns to the English cavalry. Around 1,200 Scottish prisoners were taken and estimates of the dead vary considerably. However, the initial conflict began in darkness when the English vanguard of just a few hundred came across the Scottish encampment close by Solway Moss. In the panic which ensued as the Scots believed the full English force had arrived, many of them stumbled out into the bogs and marshes and were drowned. Only six English soldiers died. Throughout history, peat and peat-bogs have been influential in conflicts and the outcomes of battles and even of wars.

In terms of heritage and archaeology of the lowland peat mosses, the two volumes by the Lancaster University Archaeological Unit, 'The Lowland Wetlands of Cumbria' (2000), and 'The Wetlands of North Lancashire' (1995), provide a remarkably detailed account of the wetlands and peat mosses of the region.

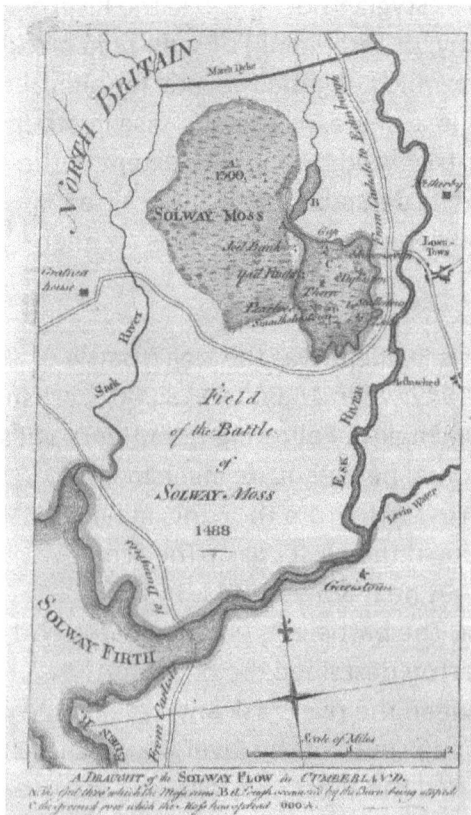

Figure 2. Map to show the site of the battle of Solway Moss which actually took place in 1542

Images and reactions to nature and landscape including peat bogs occur in the literature of travel writers and of others. So for example, in 1634 in England, the anonymous author of *A Relation of a Short Survey of Twenty-six Counties* wrote in horror of the Lake District, and that it was '...*nothing but hideous hanging hills and great pooles, that in what respect of the murmuring noyse of those great waters, and those high mountainous, tumbling, rocky hills, a man would think he were in another world.*' Daniel Defoe's description of the Peak District moors in 1724 (though often misquoted) influences reactions and expectation to this day. *'This, perhaps, is the most desolate, wild, and abandoned country in all England.'* He then goes on to state that further north in the Pennines, around Lancaster the hills were *'... high and formidable only, but they had a kind of inhospitable terror in them ....all barren and wild, of no use or advantage to man or beast.'* Westmoreland is described as a *'...country eminent only for being the wildest, most barren and frightful of any ...'* Such images help to set the scene on our attitudes on the one hand to peatlands as wastes and places of fear and loathing (Rotherham, 2013b) but on the other hand as valuable commons for the local peasants. It is these perceptions of bogs and peatlands which have influenced attitudes such as the need to wrest, recover, and restore the landscape from nature and to 'improve' the wastes so they become economically useful. The siting of the 1950s infrastructure for Britain's 'Blue Streak Missile' base, the Spadeadam Rocket Establishment, was just one manifestation of a view that these were wastes to be sacrificed in the name of progress.

Figure 3. Speed's map of Solway Moss 1611

Figure 4. Cumberland's Spadeadam Waste nr Carlisle 1962 – duct from the control room to the firing area

Figure 5. Cumberland's Spadeadam Waste nr Carlisle 1962 – the site for the Blue streak Missile

Peat & turf in and around Cumbria

Northern Cumbria: To the north of the English Lake District lies the great Solway Firth with its peat mosses, marshes, and coastal flats, and to the south are the meres, mosses, and marshes of North Lancashire. Eastwards are the expansive moors and bogs of the North Pennines and the Yorkshire Dales.

Cumbria itself has (or had) extensive valley-bottom mires and raised mires, blanket bogs and deeper peats on the fells and flatter mountain areas, and extensive marshes, bogs, and thinner peats and turf. Much of this resource has been exploited and cut, and or drained and converted to agriculture and forestry.

Ratcliffe (2002) presents a broad overview of the extent of mires and fens, and of commercial peat cutting and especially of its adverse impacts on nature conservation interests. Some of the great peat mosses such as Bowness Moss were up to fourteen metres deep. Major sites in the north around the Solway Firth include Bowness Moss, Wedholme Flow, Glasson, Drumburgh and Oulton Mosses. There are also Solway Moss, Todhills Moss, and Scaleby Moss. Inland there are extensive peatlands at Bolton Fell, and Walton-Broomhill Mosses, with other, smaller peat mosses east of Carlisle. However, the losses of bogs in this area have been considerable, with Ferguson (1892) noting that '*Scaleby, Solway and Bowness Mosses, and Wedholme Flow, are but puny and degenerate survivals of vast morasses that once covered the alluvial flats bordering on the Solway, and stretched eastwards from the vicinity of Rockcliffe along the north of Carlisle for many miles.*' Milligan (2001) highlighted the extent of reclamation on the Cardurnock peninsula with, in 1845, '*....several hundred acres of mossland known as Bowness Flow was begun to be drained.......and two years later was ready for cultivation making good farm land.*'

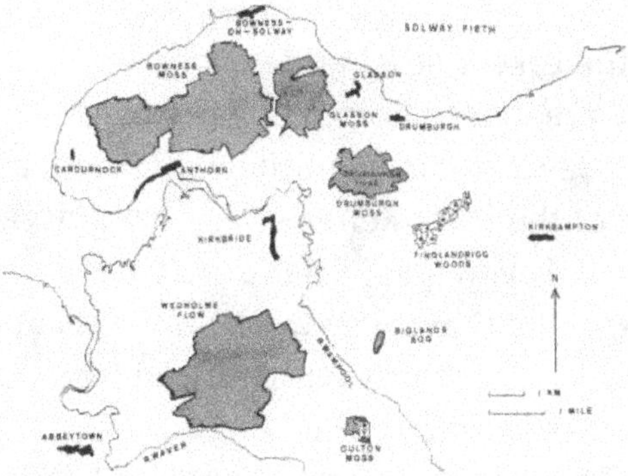

Figure 6. Boglands in the northern area

Around Cumbria - in the south: In many parts of the country peat and turf were also used on an industrial scale, sometimes in areas where today, this occurrence would seem unexpected. It may be assumed that industrial, commercial and domestic use ran alongside each other as local and regional needs demanded. David Peter in '*In and Around Silverdale*' (1984), for example, noted how, '..........*there was a plentiful supply of wood in this district to convert into charcoal for fuel. The tendency of timber supplies to become exhausted was a continuing problem for the iron producers at this period. Leighton was fortunate, however, in having an alternative fuel supply conveniently available in the peat obtained from the Arnside and Silverdale Mosses. When production was at its height, upwards of 8,000 cwts of peat a year were used at Leighton. The price of peat iron was low (ninety shillings a ton), whilst charcoal iron sold at seven pounds a ton. Occasionally quantities of peat iron had to be re-smelted.*'

Figure 7. Leighton Moss and Storrs Moss in Morecambe Bay

In the southern parts of the Lake District and beyond into North Lancashire there are, or were, raised bogs along the coastal plain around Morecambe Bay including Foulshaw, Methop, and Witherslack mosses west of the River Kent estuary. Roundsea mosses are east of the River Leven and Rusland is situated further inland. It was suggested that these were remnants of a much more extensive peatland of which Witherslack Moss was a degraded example. There are addition raised bogs either side of the Duddon Estuary with Shaw Moss, White Moss, and Herd House Moss, with White Moss at Mungrisdale and

Shouldthwaite Moss in Naddle inland. Leighton Moss and Silverdale Moss are both examples of sites with a long history of peat exploitation.

Peat bogs formerly extended along the Lake District valleys, with smaller sites around the upland tarns and blanket mires on the flatter slopes. Eastwards from Cumbria are the vast blanket mires of the north Pennines.

The main mosses began their development around 4,000 to 5,000 years ago, but some of the inland sites may be older, although in recent centuries they have been dramatically altered by human exploitation.

Cumbria also has some smaller areas of fen and fen peats though many have been drained and the remainder are now limited in extent.

The Importance of the peatland cultural landscape

The cultural exploitation of upland peatlands and commons is often neglected or misunderstood. Furthermore, before the widespread and almost total destruction of the commons, heaths, bogs and peripheral moorlands through enclosure and 'improvement', or abandonment, most local people relied on them for a diversity of products:

> **Fuel:**
>> peat
>> turf
>> ling
>> gorse/furze
>> kindling
>> birch coppice/brushwood

> **Building Materials:**
>> peat
>> turf
>> ling

stone
Sphagnum moss
bracken or fern
rush
clay
birch poles
other wood

Grazing:
sheep
cattle
ponies/horses
deer
rabbits

Other foods gathered etc.

However, the total scale of the impact of this exploitation is often not fully appreciated. The human impact on this fragile environment and on its people would be more obvious and even more dramatic if the evidence had not been almost totally removed from many areas by the sheer scale of the enclosures from around 1600 to 1800 AD. The massive conversion of heath, moor, waste, and marsh, to arable and enclosed pasture, dramatically ended this utilization of the landscape by the majority of the rural population; the rural poor and the poorer commoners. Some small-scale use of peat and turf lingered on in remote areas, particularly in the uplands, but by the late 1900s, almost all such subsistence exploitation in England and Wales had died out. Today a handful of community peat cuts occur in the Pennines and in North Yorkshire.

Figure 8. Some of the peat cutter's tools and equipment (from Rollinson, 1982)

Peat and turf were not the only fuels to be gained from heath and common. Clearly, particular use of materials varied from site to site, depending on the nature of the resource, ownership and rights, and the particular need at any one time. Gorse or furze for example was used as an alternative fuel when wood and peat were in short supply. It could also be used as kindling or for processes where a high temperature was required (such as baking).

Figure 9. Peat cutting on Witherslack early 1900s

According to Humphries and Shaughnessy (1987), the amount of gorse being brought into and stored in Dublin was so great that there were a great many complaints about the hazards of having great stacks of it near the city walls, where they *'doe overtopp the said walles in height'*. Gorse was also an important fodder for cattle and horses, particularly in winter when other crops would be in very short supply. It is likely that many commons would also have yielded some underwood and even timber, but this would not be for fuel.

Figure 10. Peat carriers at Langstrothdale, North Yorkshire 1800s

The peat resource

Since the exploitation of peat has been so widespread, involving shallow as well as deep deposits, and the peat cover has often been more-or-less completely removed, the remaining peat resource is even more diminished than previously perceived. The historical and ecological record preserved within the remaining peat deposits and buried tree remains, even if degraded, is of great value in helping understand landscape evolution. The shallower blanket peats and topogenous deposits on middle and lower level moorlands are now the scarcest resource, but there is an urgent need to protect as many peatlands as possible and to actively conserve those that are deteriorating further. On the upland moors there are now major projects to halt further deterioration of the remaining mires through blocking of drains.

The peat-cut landscape

Since peat-cutting landscapes in the uplands have remained largely unrecognised, their archaeological heritage has received no specific protection. Indeed, their protection up to now has been incidental, and due largely to the fact that land improvement over the centuries has been restricted in these upland areas by the extreme conditions and limited returns. Land uses such as sheep grazing and grouse shooting have found favour with the owners, and served to protect the historic resource to a degree; drainage being the exception.

Figure 11. Cumbrian peat ready for burning 1983

Impacts of local fuel exploitation in upland areas (cutting of peat and turf and associated drainage of vast areas of hillside and hilltop) have transformed many landscapes (Rotherham, 1998 and 1999). These impacts include the loss of upland, valley-bottom mires, and harvesting of peat and turf was compounded by extensive cutting of ling and gorse, and the gathering of kindling. Birch (*Betula* sp.) and bracken (*Pteridium aquilinum*) were extensively harvested for a variety of uses including fuel and building materials. Turf and ling (*Calluna vulgaris*) were widely used as cheap or free materials for buildings for people and livestock. These uses were described in some detail for lowland heaths by Howkins (1999) and for other areas by Rotherham (1999). The use of peat as a source of fuel for heating and cooking on home-fires is well known and extensively referenced, but most accounts are cursory and anecdotal, with the practice seldom given much weight. However, some testify to the huge scale of domestic peat cutting and its likely dramatic impact on the landscape. The scale of domestic peat cutting is especially well documented in the lowlands and less so in upland areas.

Figure 12. Witherslack Moss in the 1960s

The scale of exploitation

Much work on the cultural histories and uses of these areas has been by writers on local history such as Hartley and Ingilby (1972) in North Yorkshire and the Yorkshire Dales, but quantitative information is limited. However, in the Yorkshire Dales turf or peat has been recognised as being the chief fuel used for centuries, especially after forest clearance and even where local pit coal was available. It was cut in this area by isolated settlements, particularly at the dale heads, up to the Second World War (Hartley and Ingilby, 1968). The quantities of peat fuel consumed varied with the type and quality of the available peat, and the size of the building. In terms of the latter, consumption was affected by exposure, construction, the number and affluence of the occupants. Whether the peat was being used simply for heating or for other purposes as well, also influenced the amounts taken.

By taking local and regional accounts across the UK we can gain some insight into typical usage and hence into consumption. A farmstead with average unpretentious needs, using peat mainly for cooking and heating, would if long-standing, remove a considerable quantity of peat. For instance a traditional Cornish farm burnt about one thousand turves annually (Hamilton-Jenkin, 1932). If this was occupied for three hundred years, use about 300,000 turves (each *c.* 0.5m x 16.5cm). This is the equivalent of about 4,200 cubic metres of peat. Even greater consumption of peat is recorded for north-east Yorkshire, where the moderately sized farm of North Ghyll, Farndale, used 17,500 turves per year (Hartley and Ingilby, 1972). Since the turves cut in this Yorkshire area measured 15-18 x 5 x 4 inches (38-46 x 12.7 x 10 cm), in a 300-year period, this property would consume the equivalent of a one-metre depth area of peat 400m x 100m.

Figure 13. Typical peat-burning hearth in an old cottage

Figure 14. Witherslack Moss in the early 1900s

Upland fuel economy & the environment

Understanding the peculiarities of fuel economy in the uplands of Britain is important in informing the vision of both the upland landscape and of the communities that lived there in times past. Detailed accounts of these areas have not yet been located or interrogated but Winchester (2000) demonstrated the potential for consideration of court rolls *etc*. Other sources in the archives and in oral histories have also been noted.

Upland and northern Britain has distinctive features with respect to fuel economy:

1. High altitude is associated with longer winters and more extreme cold. Fuel consumption *per capita* may be relatively high.

2. The areas have generally been economically both backward and poor.

3. The growth rate of vegetation such as trees (and particularly woodland) may be relatively slow in comparison with lowland areas.

4. High levels of grazing stock may adversely affect vegetation and particularly woodland regeneration and resource sustainability.

5. Human impact and periodic climate change (especially long-term deterioration such as increased wetness and cold) may limit woodland recovery and may favour expansion of blanket peat at the expense of other vegetation.

6. Human occupation of these environments may be affected directly by the same climatic influences and indirectly by associated landscape change.

7. These landscapes tend to have relatively low densities of human occupation but this has fluctuated dramatically over the centuries.

8. Uplands may also be subject to transhumance – i.e. occupation only during the summer period associated with seasonal grazing.

9. Extreme environmental conditions lead to relatively comprehensive use of available natural resources. The feeding of cut holly (*Ilex aquifolium*) from upland holly hags as winter fodder for livestock is one such example. Use of gorse (*Ulex europaeus*) as fodder or fuel, and bracken for bedding are others.

Other compounding factors include remoteness and limited documentation, and perhaps restricted controls compared to lowland areas. In some regions, economically important mineral resources (lead, copper etc.) triggered intensive use and relatively high population densities albeit for limited time periods. Resulting high human population and industrial demand for local fuel sources could be significant. Factors to be considered include remote nature and frequently difficult terrain. Access to settlements was often poor, difficult, and sometimes seasonal. This affected the need to use local fuel rather than better quality materials from distant sources.

The importance of local fuel should not be under-estimated and it was often scarce. John Evelyn (1729) writing about 'fuel' noted: '……….*But besides the Dung of Beasts, and the Peat and Turf which we may find in our ouzy Lands and heathy Commons for their Chimneys, Cow-sheds, etc. they make use of Stoves, both portable and standing……… In many Places (where Fuel is Scarce) poor people spread Fern and Straw inn the Ways and Paths where Cattle dung and tread, and then clap it against a Wall till it be dry…….*'

For many commoners and other peasants, peat and turf were the main fuels. Around Milnthorpe in north Lancashire for example, peat was the main fuel until *c.*1800. An article of agreement made on 11th May 1687, between Mary Hinde of Auckenthwait Spinster and Thomas Cragge of Cragyeat in Auckenthwait, concerning '*a dwelling house and garding*', exemplifies this. It states '*that Thomas should buy the property for £8 providing that Mary Hinde shall use occupie and possess ye Chimney and ye body of ye aforesaid house during her naturall lyfe………….and that ye said Thomas Cragge shall give and lead cart to ye doore of ye said Mary Hindes house five cart full of peats or turfs in every yeare during her natural lyfe.*'(Bingham, 1987).

Figure 14. Peat cutting in Scotland in the 1700s

Examples from other regions to set the context

The scale of use is highly significant and may relate to either domestic or industrial and commercial harvesting. According to Lambert *et al.* (1961), the turf consumed by a single household for fuel, litter and other purposes, may be roughly estimated at *c.* 8,000 turves per year. In Wigtownshire in Scotland, six tons of peat was equivalent to one ton of coal. Peat stacks were as big as the cottages they heated. There are similar descriptions for North Wales up to the 1970s. At Martham in East Anglia, a household used 5,000 to 8,000 turves per year; at Scratby *c.* 10,000. This was free domestic use with no records of peat bought or sold. Peat fires traditionally burnt all year round, never going out. Required for cooking as well as for warmth, they were central to life in the uplands.

Elsewhere in Britain, commercial exploitation has been demonstrated for the Norfolk Broads, by Norwich and its Cathedral Priory, with around 400,000 turves per year used during the 1300s and 1400s (Lambert *et al.*, 1961). Similar use is described by TeBrake (1985) in Holland, and demonstrated by Ardron (1999) for parts of the Peak District and South Pennines. Significantly in the upland context, the latter showed that there was more medieval peat cut from the South Pennines (*c.* 34 million cubic metres), than from the Norfolk Broads at the same time. Most was in the early medieval period, with lower-lying sites taken during the sixteenth, seventeenth and eighteenth centuries, associated with parliamentary and private 'enclosures' of heath, moor, common, bog and 'waste'.

According to Wright (1964), when wood was scarce or at least protected in the UK, '*Peat was the only alternative fuel to be had in any quantity. There is even more of it in Britain than in Ireland, despite the poets and the travel brochures; but it was too bulky to be carried far from its source. Peat burns readily; its merit is to smoulder without a blaze, though this makes for a smoky fire. The 'peat-reek' is pleasant at a distance, and a whiff of it is not unwelcome in one's Harris tweeds; but in a 'lum' cottage its pungency is dreadful. Peat has been 'coked' to destroy the reek, powdered, mixed with pitch or rosin, and compressed into bricks that were claimed to be better than coal. Lacking true peat, some burn turf, or parings of peaty soil with roots of heather and gorse. (Peat and turf, accidentally ignited, have caused slow, widespread and unquenchable fires, which have even consumed whole villages.) Dried cowdung is a good fuel, and the scent of its fire sweeter than might be supposed; it found some favour during the wood famine, but to burn dung that ought to enrich the soil is bad husbandry.*'

Indeed, the Rev. William Harrison (in Withington, 1899) was concerned about the depletion of woods and the lack of fuel, '*Howbeit, thus much I dare affirm, that if woods go so fast to decay in the next hundred years of Grace as they have done and are like to do in this..........it is to be feared that the fenny bote, broom, turf, gall, heath, furze, brakes, whins, ling, dies, hassocks, flags, straw, sedge, reed, rush , and also seascale, will be good merchandise even in the city of London, whereunto some of them even now have gotten ready passage,........Of coal -mines we have such plenty in the north and western parts of our island as may suffice for all the realm of England; and so they do hereafter indeed, if wood be not better cherished than it is at this present........their greatest trade beginneth now to grow from the forge into the kitchen and hall, as may appear already in most cities and towns that lie about the coast, where they have but little other fuel except it be turf and hassock.'*

An advantage of peat and turf was that they were often 'free' for the cost of cutting and transportation. They were also a valuable 'free' resource to help support vital and potentially expensive services such as the schoolmaster in Scotland. The schoolmaster, '*He also has his peats cut, dried, and brought home free*'. Rev. Mr James Dingwall (*The Statistical Account of Scotland 1791-1799, Parish of Far (County of Sutherland)*)

Writing in the late 1800s Suffling (1885) described the value of peat, '*..........this is turf, and its lower brother peat. This peat, or, as it is here called, 'hovers', is, when properly dried, a capital and economical substitute for coal. It gives off a blue smoke when burning, and this, as it rises from the cottars' chimneys, wafts a rather pleasant perfume in the air, which is a great improvement on the soot-laden, evil-smelling smoke of the metropolis. A peat-ground, properly managed, is a rather valuable holding, as may be gathered from the following statistics. The peat blocks, when cut are about 4in. square (shrinking by drying to about 3¼in.) by from 2ft. to 2½ft. long (the depth of the boggy surface soil). Each square foot, therefore, produces 9 'hovers', each yard 81, each rod 2450; and, consequently, each acre the enormous number of 392,000 hovers. As these are retailed at from 1s. to 1s. 6d. per 100, a good profit must be realised.*' [This value amounts to £196 to £294 per acre and although based on a lowland site this is a useful estimate of price.]

Hartley and Ingilby in their wonderful account, '*Life and Traditions in the Moorlands of North-East Yorkshire*', stated, '*Rights of turbary appear frequently in monastic charters and peat and turf were dug in more places than would appear possible now, including the Carrs of the Vale of Pickering almost to the coast near Filey. Until the seventeenth century, fuel used at Scarborough was mostly got from the moorlands north of it and from the extensive Allerston Moor near Pickering. Special areas were sometimes allotted on the moors for the use of the tenants of a manor, and the sale of peat and turf, regulated by manor courts often caused trouble and ill feeling. From the fourteenth to the nineteenth century, records show that this occasionally took the form of the throwing down and burning of the stacks drying remote from the Company in the wide Moors.'* Even into recent times, this subsistence utilisation was an important part of the annual cycle in many upland areas. At Lockton Village in North Yorkshire this is vividly recalled by Strong (2000).

Such usage was prevalent across much of Britain and for example, the importance of peat fuel and the use of peat mosses was described in the account of Peter May, a land surveyor in north-east Scotland in the late 1700s (Adams, 1979). His diaries and accounts demonstrate the issues involved in these areas and how competing land-uses and peat extraction were administered.

Peat was extensively exploited as fuel for of iron smithing and was probably converted into charcoal before use. This application was more extensive in upland areas than the lowlands, and up to recent times highland smiths used peat charcoal. It was also a fuel for lead smelting, and in some areas in large amounts. The Yorkshire Dales lead-smelting mills are known to have used enormous quantities of peat (Hartley and Ingilby, 1968). At the Old Gang Mine in Swaledale, a year's peat was cut during late May and June and then transported to a purpose-built storage shed 391 feet long and 21 feet wide, (119m x 6.3m) (Raistrick and Jennings, 1965). In 1810, the Old Gang Mine employed 111 people in peat-getting; a well-known example of large-scale peat fuel consumption.

Elsewhere, for example on Dartmoor, peat fuel was used extensively for tin smelting and peat charcoal was the main fuel used there during the Medieval Period (Greeves, 1981). Use of peat-fuel for metal smelting was also important in and around Cumbria where it was used for copper-smelting industries. Donald (1994) indicated the scale of use during the late sixteenth century showing that for twenty weeks of operation at a Coniston smelting site required 4,800 horse-loads of peat, costing £60.

Figure 15. Carting peats by pony

Peat exploitation in & around Cumbria

Human exploitation of peat is widely recognised by authors such as Ratcliffe (2002), but this usage is mostly relatively recent and often commercially-led. He noted how cutting of peat and digging of drains had removed mosses or else caused bogs to dry. Furthermore, Ratcliffe noted how the dried out bogs became increasingly vulnerable to both deliberate and accidental fires. Drier bogs are also colonised by both heather and cotton grass, and eventually shrubs and trees.

Ratcliffe observed that *'The mosslands are one of the most important habitats surviving from a bygone age before farming dominated the lowlands'*. Wet and acidic they resisted attempts at cultivation. However, they did have great value as sources of fuel, and before coal became available and popular, Scaleby Moss for example, was almost entirely cut-over to fuel Carlisle. Similar fates befell Moorthwaite, Cumwhitton, White Moss at Walby, and probably Black Snib at Longtown. Glasson and Drumburgh Mosses were exploited in the early 1900s for a chemical works producing ammonia fertiliser, paraffin wax and other materials. Animal bones and bird guano were imported probably to mix with peat or peat extracts, though the peat may also have been used as fuel for the various processes.

The Cumberland Moss Company was established in 1948 to exploit the emerging fashion for sphagnum moss as a growing medium for gardens. They began major cutting of peat from Bowness and Glasson Mosses with peat railways and processing sheds. Soon however, they transferred their operations to Wedholme Flow. Commercial cutting in the north also occurred on Bolton Fell and on Solway Moss.

Up until relatively modern times, unless coal was available locally or from sea-ports, peat and turf would have been the main fuel for most people. Peat turbary rights were important aspects of local commons and essential for both peasant-commoners and for the landless poor. Most of the evidence for such use is relatively scattered and anecdotal, although legal disputes sometimes provide details of rights.

As described in the volume by Marshall & Davies-Shiel (1971), peat was also used as the fuel to make salt in the coastal zones, and this was an extremely important trade in medieval times. Overall impacts of such industries were probably localised and limited. However, there is more tangible information on some aspects of commercial exploitation which were far more significant. Alfred Fell (1908) wrote about the early iron industry of Furness and the surrounding areas. He noted the use of peat as fuel at Lowwood and at Leighton supplied from local mosses. At Leighton over 8,000 carts of peat-fuel were consumed in a single year. The quality of iron produced was fragile and some required re-smelting; the prices were well below that of wood charcoal-smelted iron. At Backbarrow in 1770, John Wilkinson was experimenting with the use of peat charcoal, though how successful this was is not known. He established the furnace at Wilson House near Lindale with the intention of utilizing the extensive local peat mosses. The aim was to smelt iron without charcoal and simply using natural-state peat, but this was not good enough for the process. They then tried sun-dried peats, and finally the peat compressed and dried, but again these attempts were unsuccessful. Finally, Wilkinson experimented with peat charring but still with limited success. We do have accounts of the cost of peat-winning for this work and these vary from around £3 to £1 for 'digging a double brow' and a 'head' or 'half head', and for 'One Butt digging & dressing', 3s. Another observation was that the mosses were very important in providing a water-supply to the furnaces.

Figure 16. Peat stacking at Witherslack Moss in the early 1900s

Another fascinating account is that by Collingwood in 1912, based on the accounts of the German miners in Cumbria from 1564 to 1577. This study focused on the German mining enterprises established under royal warrant, around Keswick, Coniston, and Tilberthwaite. The accounts give details of things like 'carriage' by packhorse of peat-fuel. In April 1569, the Keswick smelting operation included one man cutting up peats (6d) and six men stoking with peats (6d). Another reference was to the man who cuts up horse-loads (i.e. peat) at the smelthouses at 3s. In June 1569, the Keswick smelters were paying £214-12s-9d for the carriage of peat; a colossal sum at that time. There were also records of payment for stacking peat and measuring peat, for work at the peat stacks, and for a man with a horse to gather sods to cover the peat stacks.

In August 1569, peat carriers were paid £42 and then interestingly, a detailed comment about Skiddaw. August 18th 1569, 'Richard Scot was paid *'for a way through his ground from Schidau (Skiddaw) for one year, 8s; making that way, £1-6s-8d. Thomas Mayyson carrying 4 [long] hundred loads peat from Schidau at 2¾d.....'* There follows a long list of others paid for providing hundreds of loads of peat from a variety of locations e.g. Skiddaw, Flasco, Trylekhet, Gresdall (Mungrisdale), and High Seatt.

Figure 17. View of Skiddaw 1780

In September 1569, Christopher Mayson was again mentioned as bringing 7,200 loads at 120 the hundred and was paid £6-6s-1½d. In November, peat is mentioned as being brought from Skiddaw, Flasco, Barron Moss, Mungrisdale, and Blackcrag. Jaemes Graf of Nadell received £5 for graving a thousand peats, and John Gaiskell, foreman on Skido, was paid for 205 days at 5d a day and 45 days at 3d a day, for turning wet peats. Elsewhere, a sum of £138-11s-2d was paid for carriage of peat, and long lists of men referred to as 'peat carriers' are also given.

In 1573, £138-8s-8d was paid in advance towards a total of £221-6s-8d for *'peat delivered', and in 1584, 1,009 loads of peat at £16-16s-4d. Then, in 1575, 'John Brownrigg, bailiff of Matterdale, for his lord's peat-ground called Karl Moss, from which in 3 years (1571-74) the farmers stacked and led 43,400 loads of peat at 1s the thousand £2-3s-4d.'* In 1576, the entry reads *'Peat, 2,618 loads, £44-12s-8d'*, and in 1577, *'Peat; 3,426 loads at 4d - £57-2s-0d.'*

Figure 18. Industrial peat extraction early 1900s (not Cumbria)

The diaries sent between Cumbria and Germany end in 1577. However, the evidence presented here is of a vast industrial exploitation not only of woodlands and similar resources, but specifically of peat-grounds such as the mosses, and across the flanks of great mountains like Skiddaw. This snapshot does not provide overall detail of the amounts of peat removed but more-or-less a minimum estimate of a much greater exploitation. The implications are even more radical if we consider the associated increase in local human populations linked to the smelting and mining activities and that most of the poorer sorts of people would themselves be burning peat and turf. Wood was too valuable and the woods were protected.

Figure 19. View of Skiddaw 1800s

Figure 20. Peat ruckles in the Lyth Valley

Some specific examples of domestic peat use

Southern Cumbria & North Lancashire: The peat-cutting of this area was described in some detail by Mitchell (2005). The mosses around Lindale at the northern end of Morecambe Bay were around twenty feet deep and the old road across Foulshaw Mosses was built over bundles of faggots. The modern roads necessitated more robust engineering. Locals still had rights of turbary to cut peat-fuel on Whitbarrow Moss until well into the twentieth century, and locals recalled peat used both domestically and by blacksmiths. In particular, the smiths used a peat fire to heat the metal hoops fitted to cartwheels. Local homes were heated with peat-turves and the 'peat fireplace' at Low Levens (presumably Low Levens Farm, Levens) was described by Mitchell (2005) as having a capacity of 200 peats or turves. Peat was cut in May and dry and ready to cart by October. The cutting process began with a 'tom' spade to slice away the upper surface of vegetation, and then cutting was with a square-mouthed spade and a 'peat spade', the latter (14 inches by 5 inches) having a single wing called a 'cock'. These were used to cut the actual turves. A wooden tool with a metal blade called a 'slough' was used to clean out the drains. Cutting style might vary from site to site, and at Witherslack Moss the peat was cut horizontally. At most other sites it was cut vertically. The first layer of peat (the 'fey') was sat on top of a lower layer of 'grey peat' described as 'fuzzy'. Below this the best fuel which was the 'black peat'. There were also the dark-brown bottom-peats called 'short metal' and these were too wet and tending to break-up to be any use for burning. Finally there was a grey layer and then clay.

Figure 21. Industrial peat gas generator in Manchester late 1800s

Cut peat was lifted and the turves laid on the ground to dry prior to being set in pairs as a windrow structure of two peats on top of one another alternative layers on edge and on the flat to a height of about six or seven turves. A traditional wheel-barrow carried around fifty freshly-cut turves or 100 dried. Following the drying process the turves were stored on the peat-moss in a wooden shed with a corrugated-iron roof; being thrown in rather than stacked to aid the air circulation process. If left on the moss, then winter frost could be a problem and cause the turves to swell and break up.

Finally, the peats were transported from the moss and the drying grounds to the house or cottage which they were destined to heat. It was estimated that around 7,000 peat turves were required to heat a modest cottage for a year, and they would be carried up in loads of about 1,000 to a storage barn or stack near the dwelling. The peats were brought in for burning in pairs and each was split in half and burnt with wood, preferably ash (*Fraxinus*), and the cottage had a set of peat-bellows in order to re-start a damped fire. The fireplace had an ash-pit cleaned out each week and used as garden fertiliser. Farms such as around the Rusland Valley for example, each had a 'peat dyke' or track to lead the carts from the moss to the storage sites.

Figure 22. Industrial peat stacks (not Cumbria)

Another aspect of local culture in the peat-cutting (which is also reported from elsewhere) was that of local youths and peat-cutters smoking peat in pipes. The youths fashioned a crude bowl from an elder (*Sambucus*) stem, and pipe from a piece of honeysuckle.

The Lyth Valley & the Lakeland Dales: Gambles (1997) provides insight into this area and noted the ruined peat-huts or peat scales on the moors between Blea Tarn and Burnmoor. Situated close to sledways leading down into the valley, these huts varied in sophistication from very basic shacks to miniature barns. Most were constructed during the 1700s and 1800s, as woodland depletion and protection meant people seeking out deeper peats on the moors as a necessary fuel source. He writes of peat cutting in the Lyth Valley in the early summer with every cottage having its own 'peat-pot' scattered across the mosses and highly valued. This ownership was sometimes maintained even after widespread reclamation of the mosses. Interestingly, not only was the peat harvested for local domestic use, but also as a source of revenue when sold to households in Kendal and

Milnthorpe. Until the early 1900s, there was a regular procession of peat carts from the mosses along the road to Kendal. This ended when the canals brought cheap and easily available coal to the area. Robert Gambles describes peat as being at the heart of this rural community of the Lyth Valley with six-foot deep deposits being dug for over a thousand years and seemingly inexhaustible. Extraction increased as commercial cutting to supply local towns grew during the eighteenth and nineteenth centuries. Cottages at Causeway End were mainly homes of peat dealers, but by 1906, the local directory listed only one remaining in the Levens area.

Figure 23. Jim Whetton with an armful of Whitbarrow peat 1983

The peat-pot produced two main types of peat; one was a layer of fibrous peat frequently more than two feet deep and when dried this was used as kindling. The finer dust burnt very hot and was collected as fuel for baking bread. Below this upper layer was a depth of black peat several feet thick which was used as a general fuel, burnt well, and generated a steady heat. Following the Enclosure Acts of 1803 and 1838, the working peat-pots were allocated for commercial use by a number of individual peat dealers. They were then worked either until exhausted or until the market for peat-fuel disappeared.

Coniston: Writing in 2002, Cameron & Brown give some idea of the importance of peat to communities around Coniston, as woodland produce became valuable, scarce or protected from the 1400s onwards. Peat-fuel was then used continually from the fifteenth century up until the early nineteenth century when the Lancaster Canal brought cheap and available coal to the valley. As in other parts of England, there are accounts of peat fires such as at Lawson Park Farm burning continually for over a century. The fire never went out but would be 'damped down' overnight with a moist turf.

Farms and cottages held peat turbary rights to peat mosses on the nearby fells, with at least ten working mosses serving Coniston itself. These peat grounds were west of the village at Bleathwaite, Bannishead, The Scrow, and Stubthwaite Cove, with others on high ground over the top of Yewdale Crags. These included Kitty Crag Moss, White Crag Moss, and Yewdale Moss. The abandoned peat cuttings are still visible along with the stone platforms used to dry the peat. From the individual houses or the settlement itself ran peat-tracks to the working mosses.

Dawson (1985) describes the settlements around Torver near Coniston and the use of peat there. Trackways led from the farms and settlements to the peat mosses. Some of these were still in use within the living memories of folk in the 1980s. High Common, Back, and Low Common were both used to supply peat-fuel and the disused sites such as Throng Moss, Bullshaw Moss, and the other smaller mosses and tarns have clear evidence of abandoned peat-cuts. Once dried on the mosses and the drying grounds the peats were 'led' down to the settlements by carters. Up until the 1920s, local people still used peat rather than the increasingly available coal. Peat, cut from the local mosses, was brought down by cart or sled, and then stacked in the family peat-house and it was free or cheap.

Figure 24. Working the peat cut early 1900s (not Cumbria)

Figure 25. Old Peat hut Eskdale 1990s

Figure 26. Old Peat huts Eskdale 1990s

Conclusions

In former times peatlands were important components of the landscapes in and around Cumbria. In the past, especially influenced by human population levels and the industrial importance of Cumbria in the early industrial period, peat-fuel was extremely important. Utilisation was both commercial and industrial, and domestic, and was triggered in part by the exhaustion of wood-fuel supplies or else their protection as important resources. Until the advent of firstly canals and then railways, coal as fuel was limited by access and transport, at least anywhere beyond the immediate coastal seaport zone. However, coal was also excluded from use by the available technologies in poorer buildings which often lacked hearths and chimneys, and by cost.

Finally, beyond the nature conservation issues of late twentieth-century commercial peat extraction, the extent of peat-cutting and its impacts have been largely overlooked. The domestic use of peat-fuel has generally been ignored and the heritage or archaeology of ancient peat usage has been totally neglected. This is despite Oliver Rackham for example (pers. comm.) describing these landscapes as being at least as significant as those of ancient woodlands. The lost resource often includes the ephemeral local oral histories, peat-related traditions, and the tools and structures of exploitation.

Recent observations and discussions with local people across Cumbria have confirmed the resource of memories and cultural heritage that still waits to be recorded. Josephine Baxter for example passed on information from her uncle who farmed near Rusland and extracted peat from the moss. She also noted a farmer at Ealinghearth who also used local peat. A discussion with Caleb Brough, farmer at Midtown Farm bed and breakfast near Burgh by Sands highlighted that he formerly worked on the Solway peat mosses, and so did his father and grandfather. Their recollections are unrecorded.

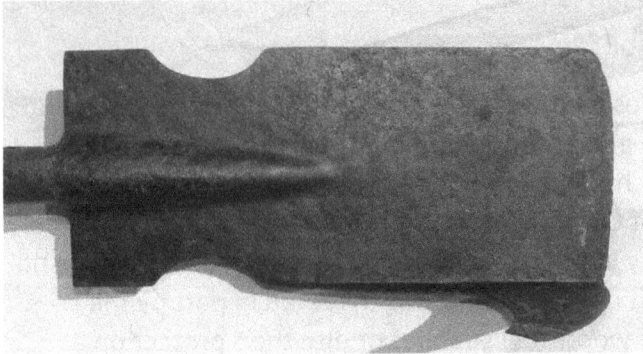

Figure 27. Winged peat spade - collection of Ian Rotherham

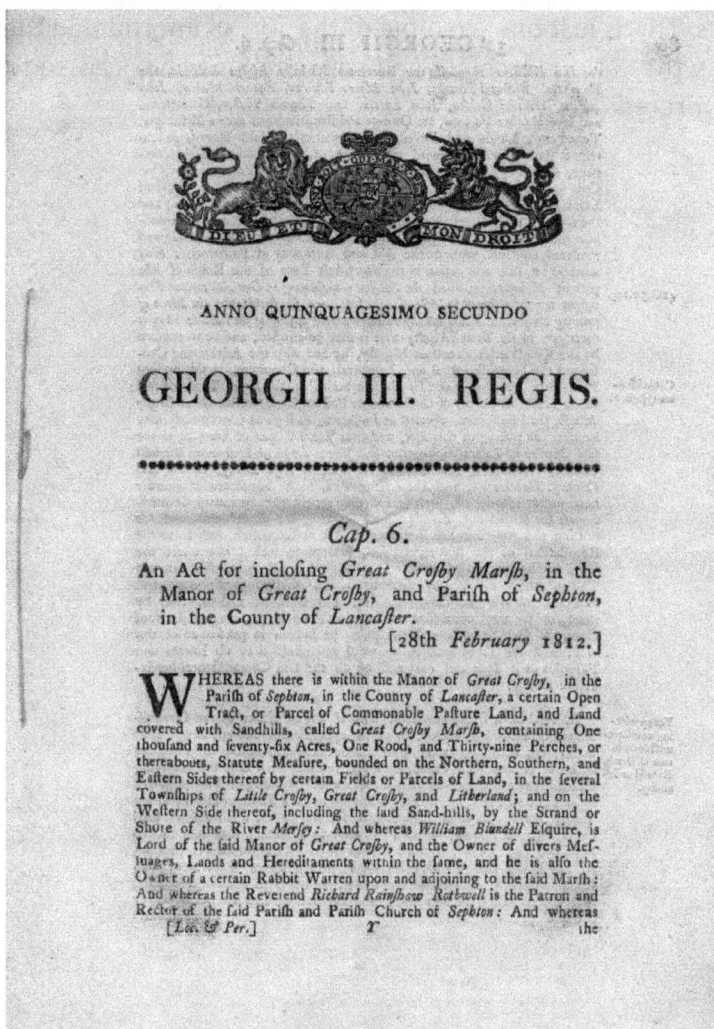

Figure 26. Act for Enclosing Crosby Marsh Lancashire 1812

Local archives also hold relevant information on peat and the peat mosses. Document No. 1/39 at the Carlisle Record Office for instance, is the 1796 award for Greystoke, Mungrisdale, Berrier, Hutton Roof and Cow End Common. Fiona Southern passed me information on the maps accompanying the enclosure award and which includes the Cumbria Wildlife Trust nature reserve at Eycott Hill east of Blencathra. Various mosses are shown on Eycott Hill (Great Aiket Pike) and appear to have rights of turbary numbered. The nearby track is shown as the 'Peat Road'. The mosses are Toll and Rum Moss, Great Aiket Moss, Cock Moss, White Moss, Garden Moss, and Larvrock

Moss. This is just one example of the sort of information that lies in the various archives and still remains to be accessed and interpreted.

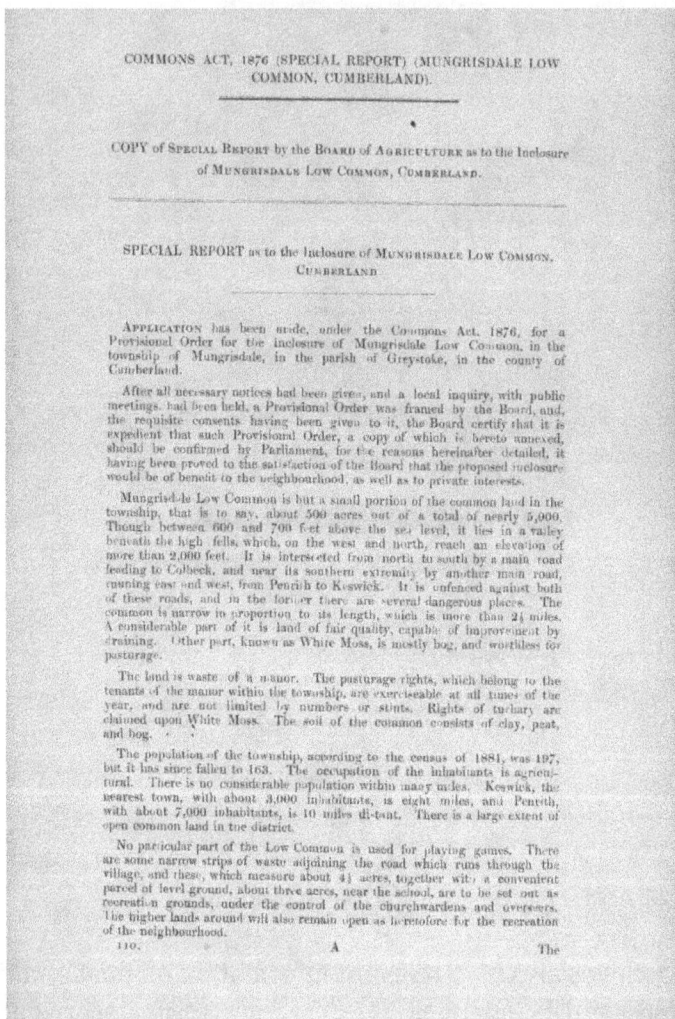

COMMONS ACT, 1876 (SPECIAL REPORT) (MUNGRISDALE LOW COMMON, CUMBERLAND).

COPY of Special Report by the Board of Agriculture as to the Inclosure of Mungrisdale Low Common, Cumberland.

SPECIAL REPORT as to the Inclosure of Mungrisdale Low Common, Cumberland

Application has been made, under the Commons Act, 1876, for a Provisional Order for the inclosure of Mungrisdale Low Common, in the township of Mungrisdale, in the parish of Greystoke, in the county of Cumberland.

After all necessary notices had been given, and a local inquiry, with public meetings, had been held, a Provisional Order was framed by the Board, and, the requisite consents having been given to it, the Board certify that it is expedient that such Provisional Order, a copy of which is hereto annexed, should be confirmed by Parliament, for the reasons hereinafter detailed, it having been proved to the satisfaction of the Board that the proposed inclosure would be of benefit to the neighbourhood, as well as to private interests.

Mungrisdale Low Common is but a small portion of the common land in the township, that is to say, about 500 acres out of a total of nearly 5,000. Though between 600 and 700 feet above the sea level, it lies in a valley beneath the high fells, which, on the west and north, reach an elevation of more than 2,000 feet. It is intersected from north to south by a main road leading to Colbeck, and near its southern extremity by another main road, running east and west, from Penrith to Keswick. It is unfenced against both of these roads, and in the former there are several dangerous places. The common is narrow in proportion to its length, which is more than 2½ miles. A considerable part of it is land of fair quality, capable of improvement by draining. Other part, known as White Moss, is mostly bog, and worthless for pasturage.

The land is waste of a manor. The pasturage rights, which belong to the tenants of the manor within the township, are exerciseable at all times of the year, and are not limited by numbers or stints. Rights of turbary are claimed upon White Moss. The soil of the common consists of clay, peat, and bog.

The population of the township, according to the census of 1881, was 197, but it has since fallen to 163. The occupation of the inhabitants is agricultural. There is no considerable population within many miles. Keswick, the nearest town, with about 3,000 inhabitants, is eight miles, and Penrith, with about 7,000 inhabitants, is 10 miles distant. There is a large extent of open common land in the district.

No particular part of the Low Common is used for playing games. There are some narrow strips of waste adjoining the road which runs through the village, and these, which measure about 4½ acres, together with a convenient parcel of level ground, about three acres, near the school, are to be set out as recreation grounds, under the control of the churchwardens and overseers. The higher lands around will also remain open as heretofore for the recreation of the neighbourhood.

110.   A   The

Figure 27. Board of Agriculture Report Mungrisdale Low Common 1891

Numerous enclosure awards and commons reports provide specific documentation of former rights, uses, and extent of peatlands. This present volume is a start but there is much still to do.

LAND COMMISSION (SPECIAL REPORT).

COPY of SPECIAL REPORT (with PROVISIONAL ORDER appended) of the LAND COMMISSIONERS as to the Regulation of DRUMBURGH COMMON and MOSS, CUMBERLAND.

SPECIAL REPORT as to the Regulation of DRUMBURGH COMMON and MOSS, CUMBERLAND.

To the Right Honourable The Secretary of State for the Home Department.

The Land Office, 3, St. James'-square. S.W.,
2 March 1885.

Sir,

WE have the honour to forward to you our Report upon an application, under the Commons Act, 1876, for the regulation of Drumburgh Common and Moss, in the township of Drumburgh, and parish of Bowness, in the county of Cumberland.

The necessary consents having been given to the Provisional Order, a copy of which is hereto annexed, we certify that it is expedient the same should be confirmed by Parliament, for the reasons hereinafter detailed, it having been proved to our satisfaction that the proposed regulation will be of benefit to the neighbourhood, as well as to private interests.

Drumburgh Common and Moss, 275 acres in extent, is about nine miles distant from, and to the west of, Carlisle, and near the Solway Firth. It is severed by the Carlisle and Silloth Branch of the North British Railway. The larger division, to the south of the railway, consists almost entirely of bog or peat moss, whereon peat is dug and stored, each person entitled to dig using a separate portion, termed a " peat pot." This part of the Common is altogether unfit for stock. The remainder, to the north of the railway, though of the same general character, with occasional patches of turf, is of better quality, and a few cows and horses are turned out upon it. The whole is so wet, even in dry seasons, that its value as pasture ground is trifling, but some improvement might be effected by draining. A foundation of sand underlies the peat moss, and there is a small quantity of clay and gravel in a few places upon the Common.

It is all waste land of a manor. The owners of lands within the township claim to be entitled to rights of pasturage, and the owners of houses to rights of turbary, and also to a right to get sand and clay for use upon their tenements. The pasturage rights, exerciseable at all times of the year, are, as already explained, of but small value. The rights of the greatest importance are those of turbary, the exercise of which is, however, a source of dispute and ill-feeling in the neighbourhood, chiefly with reference to the " peat pots," which, together with the surrounding ground to dry the peat upon, are claimed by some of the common-right owners as in the nature of private property, such claim not being acquiesced in by others.

The population of the neighbourhood is small. The village of Drumburgh has 81 inhabitants. Glasson and Easton, both townships in the parish of Bowness, and one mile from the Common on opposite sides, have, with Drumburgh township, a population of about 500. Burgh, four miles distant, has 600 or 700 inhabitants, and Kirkbride, two miles distant, has about 200. There is no populous place nearer than Carlisle, with 35,884 inhabitants, nine miles distant.

A narrow strip of the waste extends from the village, on each side of the road, for about half a mile, to the open part of the Common. From the nature of the ground, as previously described, the Common is of little or no use for recreation, nor is it resorted to or required for that purpose. On the other side of the village, a large open tract, of more than 1,400 acres, known as Burgh Marsh, stretches for about four miles along the shore of the Solway Firth. It consists of good turf, intersected by water channels, and fulfils all the requirements of a recreation ground for the neighbourhood. After regulation, and when drained, Drumburgh Common may be more fitted than it is at present as a place of resort, and those who may then be disposed to roam over it will be at liberty to do so, a privilege of enjoying air, exercise, and recreation upon the Common being reserved by the Provisional Order. There are also several other open commons, of considerable extent, in the locality.

With the exception of a few fishermen, and a few labourers employed in the manure and chemical works adjoining the Common, the inhabitants of Drumburgh are engaged in agricultural pursuits. The cottages in the township, about 10 in number, are not well provided with garden ground. Some have none at all, and the rest only very small pieces. As no part of the Common is considered suitable for garden cultivation, it has been arranged that three acres of good and accessible land near the village shall be set

120.   out

Figure 28. Land Commission Special Report Drumburgh Common and Moss Cumberland 1885

Bibliography & References

Bingham, R.K. (1987) *The Chronicles of Milnthorpe*. Cicerone Press, Milnthorpe.

Cameron, A., & Brown, E. (2002) *The Story of Coniston*. Privately published by Alistair Cameron & Elizabeth Brown, Coniston

Carnie, J.M. (2002) *At Lakeland's Heart*. Parrock Press, Windermere

Collingwood, W.G. (1912) *Elizabethan Keswick. Extracts from the original account books, 1564 – 1577, of the German Miners, in the archives of Ausberg*. Michael Moon's Bookshop, Whitehaven.

Dawson, J. (1985) *Torver: The story of a Lakeland Community*. Phillimore & Co. Ltd, Chichester

Defoe, D. (1724–1726) *A Tour Through the Whole Island of Great Britain*. Penguin, London

Denyer, S. (1991) *Traditional Buildings & Life in the Lake District*. Victor Gollanz Ltd, London.

Donald, M.B. (1955) *Elizabethan Copper: The History of the Company of Mines Royal 1568 – 1605*. Pergammon Press, London

Fell, A. (1908) *The Early Iron Industry of Furness and District*. Hume Kitchin, Ulverston

Gambles, R. (1997) *The Story of the Lakeland Dales*. Phillimore & Co. Ltd, Chichester

Hodgkinson, D., Huckerby, E., Middleton, R., & Wells, C.E. (1995) *The Lowland Wetlands of Cumbria. North West Wetlands Survey 6*. Lasted Imprints, Lancaster.

Legg, L.G.W. (ed.) (1634) *A Relation of a Short Survey of Twenty-six Counties: Observed in a Seven Weeks Journey Begun On August 11, 1634*. 1904 edition published by F.E. Robinson & co., London.

Marshall, J.D., & Davies-Shiel, M. (1971) *The Lake District at Work Past and Present*. David & Charles, Newton Abbot

Middleton, R., Wells, & C.E., Huckerby, E. (1995) *The Wetlands of North Lancashire. North West Wetlands Survey 3*. Lasted Imprints, Lancaster.

Mitchell, W.R. (2005) *Around Morecambe Bay*. Phillimore & Co. Ltd, Chichester

Ratcliffe, D. (2002) *Lakeland*. New Naturalist, Harper Collins, London

Rollinson, W. (1981) *Life and Tradition in the Lake District*. Dalesman Books, Clapham

Rollinson, W. (1996) *The Lake District Life and Traditions*. Weidenfeld & Nicolson, London

Rotherham, I.D. (ed.) (1999) *Peatland Ecology and Archaeology: management of a cultural landscape*. Wildtrack Publishing, Sheffield

Rotherham, I.D. (1999) Peat cutters and their landscapes: fundamental change in a fragile environment. *Landscape Archaeology and Ecology*, **4**, 28-51

Rotherham, I.D. (2009) *Peat and Peat Cutting*. Shire Publications, Oxford

Rotherham, I.D. (2013a) War & Peat: exploring interactions between people, human conflict, peatlands, and ecology. In: Rotherham, I.D., & Handley, C. (eds) (2013) *War & Peat*. Wildtrack Publishing, Sheffield, 7-44

Rotherham, I.D. (2013a) A Fear of Nature – Images & Perceptions of Heath, Moor, Bog & Fen in England. In: Joanaz de Melo, C., Queiroz, A.I., da Silveira, L.E., & Rotherham, I.D. (eds.) *Between the Atlantic and the Mediterranean. Responses to Climate and Weather Conditions throughout History*. Wildtrack Publishing, Sheffield, 131-162.

Rotherham, I.D., Egan, D., & Ardron, P.A. (2004) Fuel economy and the uplands: the effects of peat and turf utilisation on upland landscapes. *Society for Landscape Studies Supplementary Series*, **2**, 99-109

Rotherham, I.D., & Handley, C. (eds) (2013) *War & Peat*. Wildtrack Publishing, Sheffield

Rotherham, I.D. & McCallam, D. (2008) Peat Bogs, Marshes and Fen as disputed Landscapes in Late Eighteenth-Century France and England. In: Lyle, L. & McCallam, D. (eds.) *Histoires de la Terre: Earth Sciences and French Culture 1740-1940*. Rodopi B.V., Amsterdam & New York, 75-90

Images © Ian Rotherham

www.ukeconet.org
https://ianswalkonthewildside.wordpress.com/

Figure 29. Peat spade – collection of Ian Rotherham

Chapter 2. For peat's sake: landscapes of peat-cutting in Cumbria

André Q. Berry

Summary

This chapter provides an overview of the peat-cutting landscapes of Cumbria - historic, cultural and physical and describes some of the tools and techniques.

The Lanercost Cartulary provides probably the earliest illustration of peat cutting in the United Kingdom, certainly in Cumbria, provisionally dated to the fourteenth century. It shows peat cutters using iron-shod, heart-shaped spades, cutting 'underfoot', their foot driving the spade vertically into the substrate. Similar peat spades continued in use into the twentieth century, at least in the eastern mosslands of Cumbria. The illustrations of peat-blocks also suggest drying methods that would have been familiar to peat cutters well into the twentieth century.

A contrasting method is described from the Lyth Valley mosslands. Here the peat-cutter stood in his 'peat pot', using a 'breasting' spade driven horizontally into the face of the peat-cutting. These spades had an upstanding flange or wing on the side of the blade that enabled the peat block to be cut in two planes simultaneously.

From the late nineteenth century, diminishing demand for peat as fuel and emerging markets for peat as litter for livestock bedding and then compost for horticultural use required more systematic and effective drainage of peatlands. This was often achieved using Dutch tools and techniques.

Over 800 years of peat-cutting in Cumbria has left a legacy of features that can be read in the landscape to help better understand this once common-place activity.

Keywords: *mosslands, moss rooms, moss dales, peat, peatlands, peat cotes, peat cutting, peat scales, peat tools, turbary*

Introduction

The principal use of peat, until at least the late nineteenth century, was as fuel. It was exploited wherever it was found, because it was the only source of fuel, or because it was the only fuel that was permitted to be used, or because it was free, save for the time and labour in 'winning' it. This once 'everyday' activity was so common it was barely worthy of recording in most cases.

Probably the earliest established evidence for peat-cutting in the United Kingdom is from north-west of Balnabadoch, Isle of Barra, Outer Hebrides, where a 'fossil' pyramidal peat stack dated to the Early Bronze Age (1690 - 1490 cal BC) was found subsumed in later peat deposits (Branigan, Edwards & Merrony, 2002). Indentation marks from thumb and fingers on some of the blocks in the stack led the authors to make comparisons across the ages, between harvesting methods in the Bronze Age and the early twentieth century (*Ibid.,* p853). They also reference other evidence for the likely prehistoric use of peat (*Ibid.,* p854).

Otherwise, we must look to the Roman period for established evidence of the antiquity of peat-cutting. Irregular peat blocks found in the fill of a well at Skeldergate, York show peat was being used as fuel (Hall, Kenward & Williams, 1980), while there is evidence of extensive Roman turbaries from the English Fens, surviving as prominent parallel ridges, their silt infill now standing as positive features as the peat surrounding them is lost (e.g. Upwell (Christchurch) turbaries, TL 477 952; Figure 76, Hall & Coles, 1994). The latter suggest peat exploitation on an 'industrial' scale, to provide fuel for salt-making.

Interestingly, Domesday (1086), records only three instances of turbary, all from Lincolnshire (Darby, 1977, p352), suggesting the survey was only interested in recording peat-exploitation taking place on a significant scale or at least with financial value. That at Grainsby (347) rendered 5 solidi and 4 denarii (*toruelande v solidi et iiii denarii*), with those at North Thoresby and Autby (342b) 10 solidus (*toruelande reddens x solidus*).

It is from the medieval period onwards that we see increasing evidence of the exploitation of peat for fuel across a wide spectrum, from the domestic to the industrial scale and causing fundamental landscape change, the most celebrated of which is the creation of The Broads (Lambert *et al.*, 1960).

But what is the evidence for the 'systems' of peat exploitation in Cumbria and how is this manifested in the modern landscape?

Common of turbary

'Common of turbary' is the right to take turf or peat for burning as fuel (Cousins & Honey, 2012, p82), or for making and mending roofs or field boundaries (Dilley, 1967, p142). It is defined as a 'profit' rather than a 'customary right' in legal terms (Jessel, 1998, p60). This is because it is the right by one person to take turf or peat from the land of another and this depletes and may eventually exhaust the turf or peat resources of the land subject to the right.

The origins of the right are obscure, but as noted above significant Lincolnshire turbaries (*toruelande*) were recorded in Domesday. It must therefore pre-date the Norman Conquest. The right probably has its origins in the Anglo-Saxon period, as does the evolution of the 'manor' with which it is closely associated.

A right of turbary may extend over the whole or part of a waste depending upon the terms of the express grant (Cousins & Hey, 2012, p83) and is appurtenant, or belonging to a dominant tenement or property. The amount of turf or peat that can be cut by right is therefore limited by the number of chimneys or hearths in the property (*Ibid.*, p.101); and, cannot be used in any other property, nor can it be sold. The right could also be

45

limited by number of 'day's work' (Dilley, 1967, p143), especially relevant where the turf was for uses other than burning. An 'accident' of the Commons Registration Act 1965 was to allow some rights of turbary to be registered in gross, that is, detached from a dominant tenement or property. While Common of Turbary may not be a customary right a complex set of customs and boons or services are seen to be associated with it. This helps as a controlling exercise of the right for the common good. These matters are discussed further below in a Cumbrian context.

A Cumbrian perspective

Figure 1. Cutting peat. Todhills, Carlisle. Thomas Bushby, 1903. Dainty Series postcard in author's collection. Shows what the exercise of common of turbary may have looked like. In this case, on Rockcliffe Common, in an area now largely subsumed by the M6 motorway

The Lanercost Cartulary

The Lanercost Cartulary (CRO Carlisle, MS DZ1) provides probably the earliest illustration of peat cutters, both in Cumbria and nationally (Figure 2). Todd (1997) proposes a late fourteenth -century date for the illustrations, while recording the verdict of an expert he consulted that: '*artistic incompetence tends to be timeless*' (*Ibid.*, p41).

Incompetent or not, the illustrator appears to have had some contemporary understanding of peat-cutting. Both tools are shown with metal-shod cutting edges and their fixings; and the tools are clearly used 'underfoot' (discussed below).The drawings also appear to show the first two stages of drying the cut blocks of peat – laying them out alongside the peat bank and then, once partially dried, leaning one against another to further the drying process – stages which would have been familiar to those cutting peat in Cumbria well into the twentieth century.

Figure 2: Illustrations of peat cutters from the margins of the Lanercost Cartulary MS DZ1, folio 17 (left); MSDZ1, folio 84 (right) (CRO Carlisle, MS DZ1). Re-drawn by Timothy Morgan

These two illustrations, together with one other just showing peat-blocks (158 MS DZ1, folio 89), reference entries in the Cartulary about peat-cutting. The first (44 MS DZ1, folio 17) relates to a grant (1234 - 1256) by Alexander de Vaux to the canons and their men of Kirkland, *Lanrechaithin*, *Warthcoleman* and *Roswrageth* of common of turbary in his turbaries of Triermain (folios 16v-17). The second (215 MS DZ1, folio 84) is linked to the first and resolves a dispute over the common of turbary (1263 - 1271); while for the third (158 MS DZ1, folio 89), Ralph I de la Ferte grants the peat-bogs which belong to the saltpans which Ada Engain gave to the canons in Burgh (by Sands).

Other, entries in the Cartulary which are not illustrated also refer to peat-cutting rights (*petera*) in Kingston, Fenton and Kilmurdie, all in Dirleton, East Lothian (1100-1256, 32 MS DZ1, folio 12r-v); common of turbary for the Prior's servants and shepherds on *Brenkibet* moor (1256, 201 MS DZ1, folios 74v-76), and, other more general references to common of turbary (*et turbariis*). The latter relate to lands in Oulton (1180-1208, 155 MSDZ1, folio 48) and Alstonby (1190-1210, 88 MS DZ1, folio 30).

Turbary and the Manor

As noted above, there is a close relationship between the right of turbary and the Manor. The manorial courts (the court leets) defined the terms and places in which the right could be exercised and tried and determined the fines for those that transgressed (presentments and amercements). Dilley (1967) has analysed manorial court records for Cumbria. He notes that concerns over rights of turbary are noticeably greater in lowland manors, with four times as many presentments concerning turfs there as in the upland manors (*Ibid.*, p135). This he relates to the degree of pressure on the available resource. He also notes that the court leet exercised control over turf and peat-cutting in three ways: specifying where, when, and how much turf could be cut (*Ibid.*, p141).

Although a right of turbary may extend over the whole of the waste, as noted above, it became more usual to define areas for turf- and peat-cutting and to assign 'rooms' or 'dales' within these to individual properties or, at least, to respect the boundaries of an individual's cutting area or 'peat pot'. In Eskdale (April, 1727), for example, it was customary to keep out of *another man's peat pot* for a period of 2 years after it was last cut (C/D/Lec/94, Dilley, (1972), p.150).

The right of turbary often overlapped with a right of common of pasture. In such cases, the court leet was concerned to ensure that each right could be reasonably exercised. This it could do by excluding the right of turbary from some areas (e.g. from the out fields of Aspatria in April 1741 [Dilley (1967), p142]); or, by ensuring the turf or peat was cut in a manner that did not pose a hazard for livestock (e.g. to *dig and grave in a husbandlike manner and set the top again* [Eskdale, April 1769]; or, *shall drain the water off and shall bed the peat pots well* [Braithwaite, April 1766], both in Dilley (1967), p146). 'Setting the top again' and 'bedding the peat pots' meant putting the layer of vegetation taken from the surface before peat-cutting commenced back into the base of the cutting after the season's work was completed.

The cutting of surface turf ('topping turf' or 'flacks') for roofing ('rigging flacks') (Dilley, 1967, p142) or making and mending field-boundary walls ('spetchel', Winchester, (1994), p456) could lead to the stripping of large areas of land and a significant loss of pasture if not effectively controlled.

Turf- or peat-cutting could also undermine boundary hedgerows and walls and the court leet often set limits on how closely to such features cutting could be undertaken. This ranged from 10 yards (Wigton, April 1682) to 500 yards (Bolton, April 1760) (Dilley, (1967), p142). At Five Towns (October 1701), to prevent abuses on the common, the court leet ordered that no man should dig or grave any peats or turfs within two hundred and twenty yards of any out or head hedges belonging to Mooreside or Blindbothell - except only for repairing moor hedges and houses (Dilley, (1967), p142).

The timing of turf-cutting was often defined by the court leet with, for example, Aspatria being between 26 May and 10 June (April, 1699). At Lorton (Derwentfells) the date of cutting was at least in one instance also determined by the tenure type, with cutting after the 1st May for tenants, and after 3rd May for cottagers. There could also be variation in timing from year to year. At Wigton, the cutting period varied from between 1st and 8th June (October, 1700), to the 22nd to 31st May (April, 1708) and, from 20th May to 18th June (September, 1715) (all Dilley, (1967), p143). Dates, where specified, usually fall in May and June, but cutting in Drigg (October, 1768) was allowed from the 26th April (C/D/Pen/158; Dilley, (1972), p146)

At Westward (April, 1803), the dried turfs had to be removed from the common by 19th September (Dilley, (1967), p.146). At Staffield (September, 1637), if the peat-cutters had not led their turfs home or set them in stacks before Martinmas (11 November) each year, then it became lawful for anyone to take them for their own use '...*to the end that they may not be troublesome or hurtful to the common by rotting on the ground'* (C/D/Mus, Dilley, (1972), p151).

The quantity of peat that could be cut for burning was defined by the right of turbary as that necessary to meet the needs of the chimneys and hearths in the dominant tenement or property, as noted previously. But there was no such definition for turf cut for other purposes. The court leet effected control by setting limits on the amount of 'day's work'. At Wigton (April, 1681), tenants could have up to 4 day's work of turfs and 2 of flacks, while cottagers were allowed 1 day's work of each (Dilley, (1967), p144).

The ingenuity that some employed to circumvent these controls is also documented. At Grayrigg, in the demense land belonging to Sir James Lowther Baronet, Lord of the Manor, where there was a good peat moss or bog, some tenants had the right '......*to get as many peats as they can with one spade and others with two or three for one, two or more days, for the use and service of themselves in their respective tenements'* (CRO Carlisle, DLons 5/2/8/22 [Box 653]). A case was brought against John Fell, Gillbank in 1751 who it was alleged: '......*pretends he has a right to one spade for two days* [and] *hath for some years last past in fraud of the Custom carried with him to the peat moss in the demesne grounds ... several persons some of whom he employed in paring and clearing the surfaces of the ground that he who uses the peat spade may have nothing to hinder him, and in order to get as many cut on a day as he can, two men are set to use the peat spade by turns that the one may rest whilst the other works, and so both work more readily when at the spade, others are employed in bearing away the peats as cut, all which we apprehend should be the work of one man.'* By this means it was alleged that John Fell amassed a significant quantity of peat, surplus to the needs of his tenement, which he then sold in Kendal.

Services (or boons) attaching to a tenement and due to the Lord of the Manor for peat and turf are also recorded. For Helton Fleckett and Heltondale, surviving documents record 34 cartloads of boon peats being carted by 19 tenants on 27th and 28th July 1739. The record suggests that the service was mostly to transport the Lord's turf to his principal house (32 cartloads), but the service of one tenant, Joseph Harrison, appears also to have been to cart two loads of his own turf for the Lord's use. This is reinforced by a record of 16th June 1749, where the tenants of Helton (28 in number) transported 37 cartloads of the Lord's peats and 36 cartloads of their own peats to Lowther, 73 cartloads of boon peats in all (CRO Carlisle, D/Lons/5/2/10/52).

The complex interplay of rules, regulation, custom and services or boons surrounding the right of turbary appears to reach its apogee during the seventeenth century, as exemplified by surviving leases for the Foulshaw Mosses (Table 1) (CRO Kendal, WD/D/Misc. 1/14), part of the extensive Lyth Valley Mosslands north of the River Kent estuary. Thereafter, pressure on the peat resource apparently diminished as coal from local and regional coalfields became more cheaply and readily available; and, an individual's time and labour could be more cost-effectively deployed elsewhere, rather than in peat-cutting, drying and transporting. This occurs in parallel with the progressive decline of the manorial system.

Table 1: '*The new lease of Fowlshaw Mosses begins 26th October 1672 and ends 26th October 1683 for eleven years*' (CRO Kendal, WD/D/Misc. 1/14). Part of the original 1673 document is cut away. Missing text has been extrapolated by the author from copies of equivalent documents (dated 1718 and 1721) in the possession of the late Larry Walling, Levens [*indicated in italics within square brackets*]. There are 87 signatories.

1	First that every man that has a moss room or dale in Fowlshaw as they are now placed shall pay for every rood in breadth 1s. 4d. and so rateably, and the mosses

	to begin at Mr Thomas Biggs mosses and so down to the lower end of Fowlshaw.
2	That every man that has a moss dale or room shall pay this first year's rent the ninth day of May next after the date hereof and shall pay every year and yearly following his rent on the twentieth day of April at the now dwelling house of Thomas Ryley in Hardbreck commonly called Fishgarth House between the hours of nine and one o'clock the same day before they begin to grave or delve and so yearly during the said lease upon pain of 3s. 4d. for every default.
3	That [*every man that has a moss*] room or dale shall pay every year a day shearing if they be so demanded at the day they shall be appointed to perform the same, upon pain of 12d. for every default, and if not demanded then only 6d.
4	[*That every man or woman that has a moss dale*] or room shall for every rood in breadth lead one cart load of turves and so rateably from Fowlshaw or Intake unto Heversham Hall or to Haverbreck Hall, to Greenhead or to the Hawes [*for every two rood 1 cartful at the day they*] shall be so appointed upon pain of 3s. 4d. for every default made therein.
5	[*That every man or woman that has a moss room or*] dale shall yearly during the said lease carry from Barbon or Burton to Heversham Hall, Haverbreck Hall, Greenhead or Hawes as they shall be required two loads of coals or other service in [*lieu thereof attending as the Lord shall think fit*] in a reasonable way to be done for the same upon pain of 2s. for every default.
6	[*That no man shall grave any peats lower than the*] stakes that are now set out at the foot of their mosses and shall make a ditch across every one at the foot of his moss as shall be limited by the Lord thereof or shall be at the charge of making [*the same ditch if the lord do think it fitter*] to be done by other workmen and yearly to make good the same upon pain of 3s. 4d. for every one that makes default.

7	[*That every farmer that has a moss shall yearly*] meet and mend their ways with whins, ling and sand and to cut nothing else for the same unless they leave some sods in their own mosses where they delve their turves upon pain of 6[*s. 8d. for every default the one*] half to the Lord of Manor, the other to those that shall take the pains.
8	[*That no man being a farmer shall deny to be a Birlaw man*] for the mosses in Foulshaw once in 5 years being appointed by the Lord or his Jury at Heversham Hall and shall agree to meet and to view the mosses, and shall at any other time by the appointment of the Lord [*thereof meet to view the Mosses, and shall and will make and deliver*] to the Jury a true presentment yearly at the next Court held after Michaelmas of all and every default and defaults done against and contrary to the several articles herein set down, upon pain of 6s. 8d.
9	[*That every farmer shall be willing to agree and join together*] yearly as many as conveniently may to make one way unless they hurt the pasture too much And shall so in the same manner uphold the way upon pain of 6s. 8d.
10	[*That every man or woman that has a moss room shall yearly drain all the*] runnels belonging to his moss down to the sand, for the free passage of the water, preservation of the moss and good of the pasture before the sixth day of May next, And so continually yearly upon pain of 6s. 8d. for every default.
11	That every farmer shall every year before midsummer delve a ditch of turves up into the white moss against which time the Birlawmen agree to take a view of their mosses, and the same to be seven yards in length upon pain of 6s. 8d. for every one that make default therein.
12	That every farmer shall cleanse and make level his moss, where he spreads his turves, as also the flayings upon pain of 3s. 4d. for every default.
13	That whosoever is known to steal, take or carry away another man's turves shall for every time so doing

	forfeit 10s., whereof 5s. to the Lord of the Manor and 5s. to the party wronged.
14	That no man or woman shall remove any mear stakes or marks or boundaries between party and party upon pain of 6s. 8d. for every default.
15	That no farmer shall suffer their goods [i.e. animals/livestock] to depasture, or claim any benefit or herbage either in their own mosses or in any of the said pasture under colour of riding to work their turves or otherwise in leading them if any be known so to do to forfeit for every default 3s. 4d.
16	That every farmer shall make their appearance at the Court Leet held at Heversham Hall once every year after Michaelmas upon pain of 3s. 4d. for every default made therein, and there shall do service as other the Tenants ought to do to their Lord, by serving in the Jury when they are thereto called, and duly finding all presentments which shall be proved unto them.
17	That every farmer shall repair their moss marks yearly during the said lease upon pain of 3s. 4d.
18	That no man shall let his moss, nor take any inmates into his moss [i.e. sub-let], nor shall sell any turves upon his moss upon pain of 10s. for every default.
19	That whosoever shall be known to bring any dog, bait with dogs, or disturb the cattle, or destroy the fowl that do breed in the said grounds, shall forfeit every time so doing 6s. 8d.
20	That no man shall grave his moss twice over in one year, nor grave above eight peats in breadth upon pain [inserted: or forfeiture] of their mosses for every time so doing.

Tools and techniques

Further research is needed to properly describe the tools and techniques for peat-cutting in Cumbria, particularly for upland areas, but enough is known to provide a broad understanding. Illustrations in the Lanercost Cartulary (CRO Carlisle, MS DZ1), noted above, suggest that an iron-shod, heart-shaped wooden spade was used in the medieval period. The peat was cut 'underfoot', that is, downwards from the ground surface, using the foot to drive the spade into the peat. The resulting blocks were then spread for a period on the surrounding ground to partially dry, before being set up leaning one against the other to continue the drying process.

Research by David Park (pers. comm.) suggests that a tool of similar shape, albeit with a head of all-steel construction, appears to have continued in use into the mid-twentieth century for peat-cutting on moss lands in eastern Cumbria, including Cumwhitton Moss. On the latter, the drying process was completed by building the partially dried peat blocks into bee-hive shaped stacks some 5 to 6 feet in height, a technique also apparently used across the Lyth Valley mosslands (Figure 3, and other photographs in the Museum of Lakeland Life & Industry, Kendal collections).

Figure 3: Peat-cutting on the 'Witherslack Mosses'. (Museum of Lakeland Life & Industry, Kendal collection). The gentleman on the right is believed to be John Strickland, as shown in photographs which are dated to *around* 1935 and are on the Levens Local History Group website.

As can be seen from Figure 3, the tools and techniques of peat-cutting in the Lyth Valley mosslands differed from that apparently used across eastern Cumbria. Here the peat-cutter stood in the 'peat pot', wearing wooden 'pattens' on his feet to stop him sinking into the wet peat. He cut horizontally into the peat, usually driving the spade from chest or waist height, although the photograph shows cutting at a low level. As a consequence, these spades are usually referred to as 'breasting' spades. A second man would barrow the cut peat away from the immediate cutting area and spread it to start the drying process; or, a third man may be involved to spread the peats on the drying ground, while the second man concentrated on 'wheeling out' or barrowing the peat. While the cutting and initial spreading were usually undertaken by men, women and children were often employed in the later stages of turning and stacking the peat to dry. Women also had a significant role to play in the operation of peat businesses on the Lyth Valley mosslands.

Figure 4: This engraving by James Tingle, from an original drawing by Thomas Allom, shows loaded peat carts crossing the River Kent estuary from Foulshaw Moss (1835). Originally produced for the part-work 'Northern Tourist' series (1832-1835), London: Fisher, Son & Co.

The 1841 Census (HO1071159/14, Levens Local History Group website) lists: Ellin Jackson (65), Beathwaite Green (Sizergh Fell Side), as a *turf seller*; and, the 1851 Census, Jane Spedding (45), Levens Chapelry, as a *carter of turf* and Annie Pennie (50), Crag House, Levens Chapelry as a *turf carter*. Betsy Alderson (48), Hutton House, Levens Chapelry, is listed as a *dealer in peats* (HO107/2422, Levens Local History Group website). Presumably, these women were engaged in the peat trade to Kendal and across the River Kent estuary to Milnthorpe (Figure 4).

The late Larry Walling, Levens, was the last peat-cutter on Bellart How Moss (part of the Lyth Valley mosslands) and has described his tools and techniques of cutting (pers. comm.). His family were cutting peat on Nichols Moss in the 1930s, but started cutting peat on Bellart How Moss for their own use as fuel in 1951, when they moved to Gilpin Bridge. Larry Walling was then twenty-two years old and continued cutting each year until 2001. His family and he rented their peat pot from the Whitbarrow Estate, at a cost of £2 in 1951, rising to £10 by 2001. The 'peat pot' was located on the northern margin of the Moss.

Larry Walling's father did the peat-cutting originally (later it was Larry Walling's brother), with Larry's son, John, 'wheeling out' the peat to the drying ground where Larry laid it out to dry. Larry Walling also remembers women and children being widely involved on Stakes Moss in setting out the peat-blocks to dry. Weather permitting, peat-cutting would start in late April, but may be delayed until well into May in a bad year. The first part of the process was to remove the 'fey' or surface vegetation. This was usually done with a hay knife or ditching spade, to a depth of 1 or 2 feet depending on the nature of the surface vegetation. Tree stumps and roots were cut out with an axe. The aim was to achieve a clean and level surface for cutting. The 'fey' was thrown down onto the floor of the 'peat pot' over the previous year's cutting area.

The 'slough' (Figure 5c) was a general ditching spade, which was used to clean out drains on the Moss, particularly the drain that ran in front of the peat-cutting face. It was also used to clean the face of the peat-cutting before the season's cutting commenced, removing peat affected by frost in the previous winter. Larry Walling had a number of 'breasting' spades, some right-handed (e.g. Figure 5a) and one left-handed (Figure 5b), the latter acquired from the Holker Mosses. The 'handedness' of the spade describes the side on which the flange or wing occurs. The upturned wing meant that the spade cut in two planes simultaneously; and, the wing also helped retain the cut peat-block when the spade was being swung around to deposit the block on the waiting barrow. Although Larry Walling was right-handed he kept and used a left-handed spade when constraints on working space meant he had to turn-out the cut peat-blocks to his right-hand side.

Most of Larry Walling's tools were made for him by the Fletcher brothers, Arthur being the joiner and Joe, his twin brother being the blacksmith. There were two forges, one at Gilpin Bridge and one in Levens, at a place still known as Fletcher's Corner. The wooden patterns Arthur used to 'rough-out' handles and components for peat-barrows still survive.

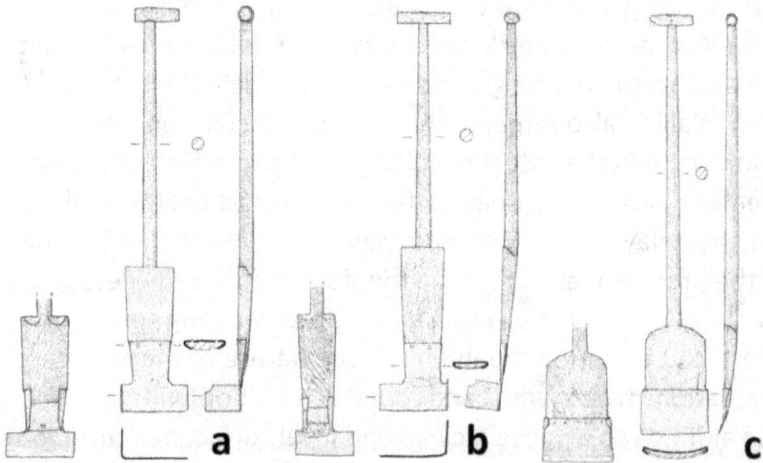

Figure 5: Peat-cutting tools used by the late Larry Walling, Levens, on Bellart How Moss. a) Right-handed peat-spade; b) Left-handed peat-spade, previously used on the Holker Mosses; c) Slough, a ditching-spade.

The nature of the peat at Bellart How Moss meant it had to be cut horizontally, standing in the cutting and driving the 'breasting' spade into the face of the 'peat pot'. If the peat was cut 'underfoot' the blocks broke up and were useless. Larry Walling's 'peat pot' was usually cut about 6 feet wide and was 6 feet in depth, down to a layer of grey clay. The peat-spade cut blocks approximately 1 foot long and 4 inches on each side. If the going was good, 1,000 blocks could be cut and set out to dry by 3 men in 3 hours. Where the cutting was difficult, this may fall to 400 blocks. The family would use about 15,000 blocks each year, that is, about 45 cubic metres.

Occasionally, the peat-cutter would find a 'sheck', a hidden crevice in the peat full of water. Where cutters were unlucky this could release *hundreds of gallons of water*, temporarily flooding out the cutting and causing the cutter to have to move to a higher level to continue working. The cut blocks were deposited straight onto the peat barrow which, when full, was 'wheeled out' to the drying ground, usually set some 20-30 yards away from the cutting face on the cut-over surface. This distance may be as much as 50 yards if it was intended to cut a large number of blocks in a season.

In the early days, the blocks were often set out to dry in rough 'windrows', loosely piled some 8 or 9 blocks in height while still wet, with gaps between the blocks to allow the wind to pass through. They would then be moved into 'stools' to continue the drying process, two blocks side-by-side with a gap between, with two blocks set on top of them at right angles, this repeated until there were 5 or 6 layers some 2 feet in height. Blocks would also be set out wet in a single layer, flat in rows, in what Larry Walling's father called the 'Witherslack way'. Only when partially dried would the blocks then be set up in 'stools'. It was this method that Larry Walling used in later years. When dried, which could be July in a good year and October in a bad year, the peat-blocks were stored in corrugated iron-clad sheds on the Moss until required for burning, when they were transported by tractor and trailer. The difference in tools and techniques between the Lyth Valley and other mosslands and

Cumwhitton Moss and the eastern mosslands of Cumbria requires further research.

In the Fens of eastern England, Lucas (1930, p23) notes that a heart-shaped, *moorland* spade was used for 'underfoot' peat cutting up until 1856, when it was replaced by the fenland *becket*, first introduced to Isleham Fen. The moorland spade is similar in appearance to the spades used for peat cutting in eastern Cumbria, while the becket is similar to the 'breasting' spade used on the Lyth Valley mosslands, but was used 'underfoot'.

What accounts for the different tools and techniques across Cumbria, and of what antiquity are they? Are they only a response to the different types of peat and how it has accumulated, or are there cultural factors at play?

Mosslands -what are they good for?

As the importance of peat fuel diminished from the nineteenth century onwards with the availability of cheap alternative fuels, principally coal, developments in agriculture and the industrial revolution created new demands for peat. These further increased pressure on the peatland resource.

With the advent of 'scientific farming' in the nineteenth century, there was both an economic and cultural drive to bring marginal lands and 'waste' into productive agricultural use to feed an increasingly industrialised and urbanised nation. The landed classes considered it almost a duty to 'win' land from the waste and much was written about ways in which to do this (e.g. Smith, 1856), with large sums and ambitious techniques employed in an attempt, often futile, to do so (Bonnett, 1965; Hegarty & Wilson-North, 2014). Peat-cutting for fuel often aided and abetted this process. As the peat was cut from the margins of the bog in towards its centre, so the land behind could be 'clayed' and first brought into pastoral use (Larry Walling, pers. comm.) and then, as drainage improved, brought into arable rotation. Enclosure further assisted this process. On the Lyth

Valley mosslands, enclosure has allowed the mosses to be brought into productive agricultural use, albeit sustained by continual pumped drainage. Within this mossland complex, on Quaggs Moss, the 'moss ends', surviving baulks of uncut peat along the inward-margin of the individual moss 'dales' or 'rooms', can still be seen, preserved in woodland clumps. These help demonstrate the role of peat-cutting in the 'improvement' process.

In an age before petroleum products, there was extensive interest in the destructive distillation of peat to yield, paraffin, napthalene, gas and other such products, for heating, lighting and industrial uses (Kerr, 1905). Companies came and went, with financial support from speculators in London and elsewhere, such companies often employing the latest patented technologies for peat-winning and processing (Kerr, 1905; Berry *et al.*, 1996, pp87-104.). Rotherham (2011) explains the many and varied uses for peat. However, there are two uses for peat which have wrought a fundamental change to the lowland peatlands of Cumbria and the UK as a whole – as bedding for livestock, and then as a growing medium for horticulture.

Before the advent of the internal combustion engine, motive power in an increasingly industrialised and urbanised society was provided by horses, which needed bedding down and mucking out. 'Peat litter' was found to have many advantages as a bedding material – it was insulating, could absorb large quantities of liquids, and had sterilising and odour-neutralising properties (Kerr, 1905). Animals also did not tend to eat it (although in other circumstances peat was mixed with molasses to create a feed for cattle (Björling & Gissing, 1907)). Peat was also used in poultry housing for the same reasons (Berry *et al.*, 1996). Upon disposal, the mix of peat litter and animal urine / dung made a good soil improver and fertiliser. In this context, the Lound Coal & Moss Litter Company (proprietor, Haley Wormald) is recorded as having an office and wharf at Old Laund, Kendal on the Lancaster Canal in 1894 (Kelly) from where it supplied peat litter.

Peat has been used by specialist nurseries as a growing medium since at least the late-nineteenth century (Kerr, 1905; Johnson, 1866). Nevertheless, it was not until after the Second World War, promoted by horticulturalists such as Percy Thrower, that it became universal in use by both professional growers and amateur gardeners (Berry *et al.*, 1996).

These new industrial and domestic uses brought with them a significant increase in demand for peat and the need for a more systematic and effective drainage of peatlands.

Going Dutch

The Dutch drainage engineers and workers were hugely important in the drainage of the English Fens in both East Anglia and in the Humber Levels (Rotherham, 2010, 2013). Dutch peat-cutters and working methods are first definitively recorded from the Humber peatlands in 1894, although may have been present from as early as 1887 (Limbert, 1986). The involvement of the Griendtsveen Moss Litter Co., a Dutch company, from 1894 establishes the link with Cumbria.

Richardsons' Moss Litter Co. had acquired a lease on Solway Moss in 1896, having been founded as the Union Moss Litter Co. by George Gibson Richardson, at the age of 17, in Sandhill, Newcastle upon Tyne. Richardson's merged with the Klazienaveen Moss Litter of Groningen in 1904 to form the London & Provincial Moss Litter Co. Ltd., going on to join the Peat Moss Litter Supply Co. in 1906, of which the Griendtsveen Moss Litter Co. had been a founder member (L&P Peat Ltd, 1983[?]) Such is the complex world of peat companies in the United Kingdom.

By this means it seems that Dutch workers, tools and techniques arrived in Cumbria; and from there were to go on to introduce the Dutch method to Fenn's and Whixall Mosses on the Shropshire / Wrexham County Borough border in the 1920s (Berry *et al.*,1996).

The first operation was to open up a grid of causeways ('caseys') and drainage channels across the peatland, between which were the cutting flats, with the 'benches' where cutting took place being set at right angles to the ditches in the flats (Figure 6; Berry *et al*, 1996).

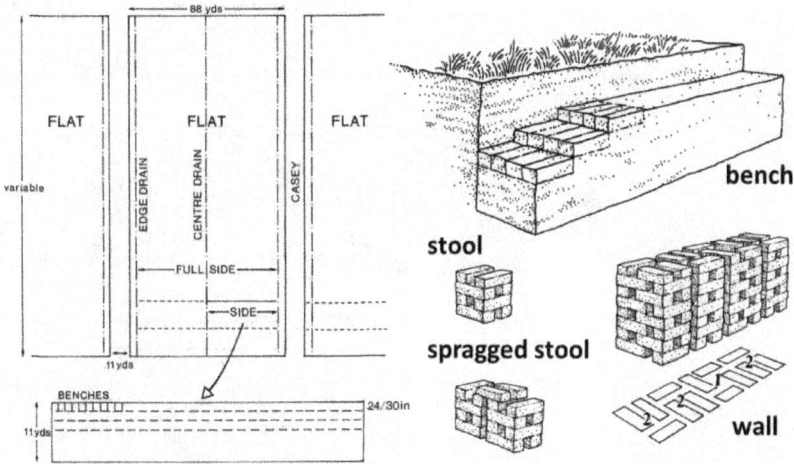

Figure 6: The Dutch method of peat cutting, showing the disposition of drains, causeways ('caseys'), cutting flats and benches; and the three stages of the drying process, stools, spragged stools and walls. From Berry *et al*, 1996. Drawings: Timothy Morgan.

The surface vegetation was first removed from the bench using a 'feyer' or similar spade to create a clean, level surface (Figure 7a), the 'fey' being thrown down into the bottom of the cutting area. The 'nicker out' (Figure 7b) was then used to mark out the edge of the bench, before the 'sticker' (Figure 7c) was employed to cut the ends and width of the individual peat blocks. Standing in the cutting, the peat-cutter then used a 'breasting' spade, variously called the 'bat', 'uplifter', or 'oplegger' to make the final cut of the underside of the individual peat block and lift and deposit it onto the bench-top alongside the cutting. It is to be noted that the bat does not have an upturned flange or wing, because the cutting was a two-stage process, with the side cuts of the blocks first made with the 'sticker'.

The cut blocks measured *c.*16 inches long and 6 inches on a side. On Fenn's Moss, peat-cutters were paid piece rate, by the 'Dutch chain', a measurement of 44 cubic imperial yards or 3,564 peat blocks. An experienced cutter would expect to cut a Dutch chain a day. In the 1920s, the rate was 12s-3d per 'Dutch chain', rising to 42s by 1952 (Berry *et al*, 1996, p107).

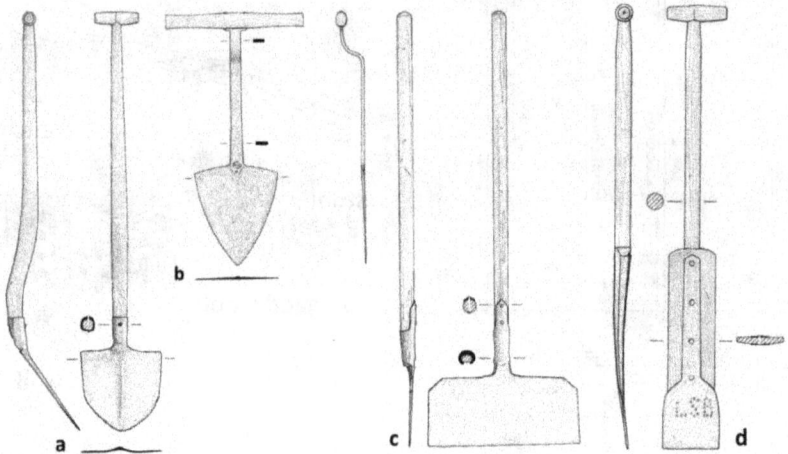

Figure 7: Dutch method tools. The tools illustrated are from Fenn's and Whixall Mosses (from Berry *et al*, 1996), but are very similar to tools used on the Cumbria peatlands. Drawings: Timothy Morgan. a) 'Feyer'; b) 'Nicker out'; c) 'Sticker'; d) 'Bat', 'uplifter', or 'oplegger'.

After an initial period of drying, the blocks would be stacked into 'stools' (Figure 5) to further assist the drying process and may then be reconfigured into 'spragged stools', moving the wetter, basal peat blocks higher up in the stools. The final stage was to build the stools into 'walls'. Once dry, the walls were usually 'wheeled out' by peat barrow to the causeways, where they were built into enormous stacks until required at the peat-mill. Hand-cutting using the Dutch method continued into the late 1960s on most peatlands until machines arrived from Germany that more-or-less replicated the hand-cutting process.

The landscapes of peat-cutting

So what impact has at least 800 years of peat cutting had on the landscape of Cumbria; and, what evidence remains of this once domestic and ubiquitous activity? By definition, at least to a certain extent, peat-cutting is self-erasing, the resource is cut away and may become exhausted; or, it facilitates agricultural reclamation and improvement that itself destroys the evidence. An example of the latter has already been noted on Quaggs Moss in the Lyth Valley mosslands, where the remains of 'moss ends' are the only survival of a once extensive peat cutting landscape, now preserved in small woodlands across the enclosure landscape. These survive as upstanding baulks of eroding peat, remnants of the one-time cutting faces, now subsumed beneath trees, shrubs and other woodland vegetation.

In the uplands of Cumbria, too, areas of farm or community peat-cutting are becoming lost beneath heather and moor grass, the cutting-faces progressively eroding and softening with each season's frost. It is likely that many an eroding peat hag is, in fact, an abandoned peat cutting, rather than a natural feature, particularly where these occur in clusters. But peat-cutting has left its mark on the Cumbrian landscape in a more tangible way.

The thirty-five 'peat scales' or storage huts of Eskdale are well-known, clustered above Boot and connected to that community by a zigzag sled track and also distributed around the boundary of the valley between fell edge and valley scarp, where they served individual farms (Figure 8) (Winchester, 1984). They appear to date from the sixteenth to nineteenth centuries and Winchester suggests that they are a response to the difficulties of 'winning' peat during the summer months, allowing for partially dried blocks to be removed to storage to complete the drying process under cover.

Figure 8: 'Peat scale', Boot, Eskdale

Winchester (2000, p131) also notes 'peat scales' in Langdale, Bampton, Barton, Dunnerdale and Longsleddale. There is also a 'peat scale' above Swinside, its function helpfully confirmed by the adjoining Peathouse Beck. What appears to make the 'peat scales' of Eskdale particularly noteworthy is their grouping above Boot. But this may not be unique. 'Peat scales' are often referred to in historic documents as 'peat cotes' and work by Levens Local History Group (pers. comm.) suggests that the community of Cotes, north of Levens, on higher ground on the margin of the Lyth Valley mosslands, comprises a similar cluster of 'peat scales' or 'peat cotes', now largely converted to garages or dwellings. Other evidence of peat cutting is revealed by patterns in the landscape – parallel boundaries or boundaries converging towards the centre of a moss or a single point, that indicate the past existence of moss 'dales' or 'rooms'. Some evidence is best seen from the air (Figure 9).

Domestic-scale cutting was usually undertaken around the margins of the mosslands, because of the difficulties of drainage. Evidence of this is manifest as ragged edges that can be seen around many of the lowland Cumbrian mosslands. Dutch cutting patterns can also still be clearly seen on Glasson Moss, Bowness Common and Wedholme Flow (aka Kirkbride Moss), but have been lost from Solway Moss, where more recent levelling and surface-milling of the peat to provide compost for horticultural use has erased the earlier patterns of cutting.

Figure 9. Nutberry Moss, north of the Solway, *c.*1967, looking east. By kind permission of I.G. Richardson. Domestic-scale peat cutting incursions around the margins of the Moss are clearly visible, as is the Dutch cutting pattern. Similar patterns can be seen on mosses on the Cumbrian side of the Solway

Conclusions

This chapter provides an overview of the historic, cultural and physical aspects of peat-cutting landscapes of Cumbria. However, more research is needed to properly understand the tools, techniques and evidence of peat cutting, particularly in the uplands of Cumbria.

Nevertheless, enough of a story can be told to reveal the legacy of this once every-day, ubiquitous activity, so common that it was often considered barely worthy of recording in most cases.

Acknowledgements

Thanks are due to the staff of Cumbria County Council Archives Service at Kendal and Carlisle; to David Park and to members of the Levens History Society (particularly Peter and Gillian Wood and the late Larry Walling) who gave freely of their knowledge and expertise; this chapter could not have been written without them. Thanks are due also to Natural England, for funding the illustration of peat tools and the author's time in carrying out some of the research for this paper; and to Timothy Morgan for the illustrations of peat tools and techniques.

References

Berry, A.Q., Gale, F., Daniels, J.L., & Allmark, W. (1996) *Fenn's and Whixall Mosses*. Clwyd County Council, Mold.

Björling, P.R., & Gissing, F.T. (1907) *Peat: its use and manufacture*. Charles Griffith & Co., London.

Bonnett, H. (1965) *The Saga of the Steam Plough*. George Allen & Unwin, London.

Branigan, K., Edwards, K.J., & Merrony, C. (2002) Bronze Age fuel: the oldest direct evidence for deep peat cutting and stack construction? *Antiquity*, **76**, 849-855.

Cousins, E.F., & Honey, R. (2012) *Gadsden on Commons and Greens*. 2nd Edition. Sweet & Maxwell, London.

Darby, H.C. (1977) *Domesday England*. Cambridge University Press, Cambridge.

Dilley, R.S. (1967) The Cumberland court leet and use of the common lands. *Transactions of Cumberland and Westmorland Antiquarian and Archaeological Society*, **67**, 125-151.

Dilley, R.S. (1972) *Common Land in Cumberland, 1500-1850*, unpublished thesis, University of Cambridge.

Hall, A.R., Kenward, H.K., & Williams, D. (1980) *Environmental evidence from Roman deposits in Skeldergate*. In: Addyman, P.V. (ed.) The Archaeology of York, Vol.**14:3** *The past environment of York*. Council for British Archaeology, London.

Hall, D., & Coles, J. (1997) *Fenland survey: an essay in landscape and persistence.* English Heritage, Archaeological Report 1. English Heritage, London.

Hegarty, C., & Wilson-North, R. (2014) *The Archaeology of Hill Farming on Exmoor.* English Heritage, Swindon.

Jessel, C. (1998) *The Law of the Manor.* Barry Rose Law Publishers Ltd., Chichester.

Johnson, S.W. (1866) *Peat and its uses, as fertiliser and fuel.* Orange Judd & Co., New York.

Kelly & Co. (1894) *Kelly's Directory of Cumberland & Westmorland,* Kelly & Co., London.

Kerr, W.A. (1905) *Peat and its products.* Begg, Kennedy & Elder, Glasgow.

L&P Peat Ltd. (undated, 1983?) *100 years' progress in peat, 1883-1983.* Promotional leaflet. L&P Peat Ltd., Carlisle.

Lambert, J.M., Jennings, J.N., Smith, C.T., Green, C., & Hutchinson J.N. (1960) *The making of The Broads: a reconsideration of their origin in the light of new evidence.* R.G.S. Research Series No. 3, The Royal Geographical Society, London.

Limbert, M. (1986) The exploitation of peat at Thorne. *Old West Riding,* **6**, No.1. Old West Riding Books, Huddersfield.

Lucas, C. (1930) *The Fenman's World: memories of a Fenland physician.* Jarrold & Sons, London.

Rotherham, I.D. (2010) *Yorkshire's Forgotten Fenlands.* Pen & Sword Books Limited, Barnsley

Rotherham, I.D. (2011) *Peat and Peat Cutting*. Shire Publications, Oxford.

Rotherham, I.D. (2013) *The Lost Fens: England's Greatest Ecological Disaster*. The History Press, Stroud.

Smith, R. (1856) Bringing Moorland into cultivation. *Journal of the Royal Agricultural Society of England*, **17**, 349-394.

Todd, J.M. (ed.) (1997) *The Lanercost Cartulary*. Surtees Society Vol.203/ Cumberland & Westmorland Antiquarian & Archaeological Society, Record Series Vol.XI.

Winchester, A.J.L. (1984) Peat storage huts in Eskdale. *Transactions of Cumberland and Westmorland Antiquarian and Archaeological Society*, **84**, 103-115.

Winchester, A.J.L. (ed.) (1994) *The Diary of Isaac Fletcher of Underwood, Cumberland 1756-1781*. The Cumberland and Westmorland Antiquarian and Archaeological Society, Extra Series XXVII.

Winchester, A.J.L. (2001) *The Harvest of the Hills: rural life in Northern England and the Scottish Borders, 1400-1700*. Edinburgh University Press, Edinburgh.

Websites

Levens Local History Group, Old Photographs, Peat Cutting (accessed 6 December 2018):
http://www.levenshistory.co.uk/ph_peat.html#peat

Levens Local History Group, 1841 Census (accessed 6 December 2018):
http://www.levenshistory.co.uk/census/1841%20census.pdf

Levens Local History Group, 1851 Census (accessed 6 December 2018):
http://www.levenshistory.co.uk/census/1851%20census.pdf

Chapter 3. Historical change in, and future prospects for, the raised bogs of Cumbria

Richard Lindsay
Sustainability Research Institute, University of East London

Summary

Analysis of historical land-use data has shown that lowland raised bogs in Britain have undergone enormous change during the last 150 years. The study has revealed some startling figures concerning change, suggesting that losses of this habitat since the mid-1800s have amounted to more than 90% in terms of the remaining near-natural areas of lowland raised bog habitat. More extensive subsequent survey has confirmed this scale of habitat degradation and loss, highlighting the fact that surviving near-natural habitat consisted largely of partial sites or fragmentary remnants. The implications of this for the long-term conservation of raised bog habitat are profound, especially when linked to the hydro-morphological controls implicit in the 'ground water mound theory' when applied to raised bog systems. Indeed, such implications not only represent major challenges for raised bog conservation but also have global significance, given the current state of lowland raised bogs around the globe. Recent developments in thinking about land-use on lowland peat soils, however, now offer considerable promise in terms of reducing or even reversing these challenges as part of a revolution in agricultural perception and practice based on the concept of 'paludiculture'.

Keywords: raised bog, hydrology, morphology, paludiculture, peatland condition

Introduction

The north-west of England has long been known as a stronghold of lowland raised bogs but various events during the twentieth century was giving rise to a growing sense of unease about the future of this habitat. In 1976, a study was therefore initiated with the aim of putting the state of lowland raised bogs in Britain into a context of historic land-use change associated with raised bog systems (Bragg *et al.*, 1984). Acquisition of such data was increasingly thought to be urgently needed in order to underpin arguments for conservation of surviving raised bog habitat. This was given the development pressures that were demonstrably and increasingly being applied to this habitat, particularly from afforestation and commercial peat extraction. The habitat had already been the subject of some extremely contentious and high-profile conservation arguments in relation to planning consents for commercial peat extraction in northern England but the rapid expansion of conifer plantations across many raised bogs, particularly in Scotland, was also giving rise to considerable concern.

Historical survey of lowland raised bogs in Britain – the Cumbria picture

Bragg *et al.* (1984) focused on four concentrations of lowland raised bogs distributed across northern England and the Scottish Lowlands, namely the Lancashire lowland plain, the south Cumbrian river valleys, the English and Scottish shores of the Solway Firth, and the Forth Valley. Five sources and dating periods were identified that would allow for consistent mapping. For Cumbria, three groups of sites were identified in south Cumbria, namely those in the Duddon Valley, the Leven Valley and the Kent Valley, and two in north Cumbria, specifically sites around the River Whampool. There was also a group of sites to the north of Carlisle termed the 'Gretna' sites.

Four categories of condition class were defined for all sites across the whole survey based on what could reliably and consistently be identified from the five mapping sources used, namely 'open mire', 'drained mire', 'wooded mire' and 'agricultural land-claim'. Widespread loss of bog habitat to agricultural land-claim emerged during the latter part of the nineteenth century with only modest losses to this activity until the post-war period. At that time, increased mechanisation and the availability of land drainage grants encouraged further land-take. Against this background, the pattern of land-use change differed somewhat between the bogs of north and south Cumbria. Industrial peat extraction during the inter-war and post-war years resulted in the greatest loss of habitat from the bogs of north Cumbria whereas the picture was more mixed in the south Cumbria valleys. Following the land-take of the nineteenth century which affected all three valleys equally, domestic peat cutting was the major activity in the Duddon Valley. This was until the 1970s when agricultural land-take again became a major factor. In the Leven Valley and Kent Valley, losses occurred to domestic peat cutting, agricultural land-take and encroachment of tree cover across drying bog expanses. There was also wholesale planting of alien conifer species across the largest site in south Cumbria (Foulshaw Moss) as a multi-species and multi-treatment afforestation experiment undertaken by the Forestry Commission.

Overall, it was found that by 1978 the Whampool group of north Cumbria had retained the highest proportion of open bog habitat (i.e. not cut, afforested or drained). In 1978, these sites still supported 768 ha of their original 2,172 ha (35%), but because losses had been so severe in other parts of the county, this also represented 81% of the surviving open bog habitat in Cumbria. Most of the remaining bog area lay either in the Duddon Valley (72 ha) or the Kent Valley (53 ha), largely scattered across the three valleys as small patches within otherwise damaged sites.

However, survival as open bog habitat was not the whole story, because many such areas have been subject to fire damage, peripheral drainage and other factors inimical to actively peat-forming vegetation. Bragg *et al.* (1984) provide values for the quality of this remaining open bog habitat across the four main study areas, and in doing so, highlight that only some 25% to 26% of the open bog could be classed as high-quality (i.e. near natural) bog vegetation. The remainder was of lesser, or poor, quality and often no longer supported a vegetation which is normally regarded as peat-forming.

Placing the data obtained for Cumbria into the context of the wider study, these bogs originally represented 31% (4,469 ha) of the total area of bog mapped across the four study areas (14,257 ha). Yet by 1978 the Cumbrian sites contained 53% (952 ha) of the total remaining open bog area (1,803 ha) recorded across the entire study. As such, Cumbria now represented the main stronghold for surviving open bog within the overall study. The bulk of the remainder was provided by two sites in Scotland – East Flanders Moss in the Forth Valley (548 ha) and Longbridge Muir within the Lochar Mosses complex of the north Solway bogs (190 ha). Not only do these open areas of bog in Cumbria represent the most extensive surviving remnants of such habitat, they also provide the most extensive examples of high-quality peat-forming habitat within the entire study area. Some 250 ha of highest-quality bog habitat survives in the Cumbria sites. This is twice the size of the next-largest area of high-quality habitat, provided by Longbridge Muir on the north Solway shore.

From the study of four concentrations of raised bogs in lowland England and Wales, therefore, it emerged that 87% of the original extent of raised bog habitat present in the mid-nineteenth century no longer supported open bog habitat. Furthermore, of the 13% of such habitat that remained, only half was in fair to good condition, thus representing 7% survival and 93% loss of near-natural raised bog habitat between the 1850s and 1978. Cumbria therefore provided a key refuge for such habitat (see Figure 1).

It is important to highlight, which many have failed to do, that the area of surviving peat bog habitat is significantly greater than the 7% of *near-natural* bog habitat indicated here. This is because extensive areas of damaged and highly degraded bog remain. These are areas which have, for example, been subject to domestic peat extraction or which have been intensively drained. These damaged areas of raised bog amounted to 2,819 ha or some 20% of the original extent of raised bog habitat.

Inventory of all lowland raised bogs in Great Britain – the contribution of Cumbria

Following on from the sample survey reported on by Bragg *et al.* (1984), a complete inventory of peatlands in Britain was undertaken by Lindsay & Immirzi (1996). Their survey included a particular focus on the condition of lowland raised bogs and was based on the extent of peat mapped by the British Geological Survey (BGS) as 'peat' on their 1:50,000 maps of superficial deposits. 'Peat' was only recorded on these BGS maps when it was at least 1 m deep, which had the benefit in the lowlands of therefore generally only mapping raised bog peat deposits. Site types were also confirmed by correlation with peat soil types identified by Burton & Hodgson (1987) as part of the Soil Survey of England and Wales.

The BGS superficial drift maps were taken to represent a minimum extent of original lowland raised bog across the three countries. This was a minimum extent, because it was recognised that many raised bogs had been subject to such intensive exploitation around their margins that in some cases the raised bog peat soil had been completely lost even before BGS field surveys in the 1870s. Assessment of site condition drew on actual field surveys, most notably Bragg *et al.* (1984) for a range of sites across England and Scotland, and McTeague and Watson (1991) for sites in the Central Belt of Scotland, together with aerial-photographic assessment from the collection held by Scottish Natural Heritage, and consultation with staff in the three country conservation agencies.

Lindsay & Immirzi (1996) identified 69,663 ha of lowland raised bog peat soil in Britain, which they took to be the minimum original area of raised bog habitat prior to extensive land-take by various forms of land use. On categorising the areas of identified raised bog peat soil, Lindsay & Immirzi (1996) found that, while there were regional differences, the overall pattern of land-use change identified by Bragg *et al.* (1984) was repeated nationally. Despite a hitherto unrecognised concentration of raised bogs in the Central Belt of Scotland, the overall pattern of identified change revealed that the area of 'primary near-natural raised bog habitat' remaining in 1995 from the original 69,663 ha amounted to 3,836 ha – representing just 6% of the original extent. Of this, Cumbria contributed 448 ha, representing some 12% of this surviving resource.

Damaged margins – a common feature of raised bog systems

While the inventory undertaken by Lindsay & Immirzi (1996) identified a significant area of raised bog habitat still in near-natural condition, it must be emphasised that this consisted almost entirely of relatively small portions of once-larger sites. Not a single site remained entirely near-natural. Every raised bog identified in the inventory as still retaining some open bog habitat had suffered some degree of damage to the bog margin. Furthermore, every site listed in the inventory had lost its marginal lagg fen zone – that is, the transition zone between bog peat and adjacent ground.

This fragmentary nature of the surviving open raised bog remnants, combined with the loss of surrounding lagg fen, has profound implications for the long-term character of these remnants and the biodiversity of lowland raised bog habitat as a whole in the UK. Indeed, the problem is not restricted to the UK. A great many national inventories across Europe and further afield have identified that many surviving expanses of raised bog habitat now represent only part, or parts, of their original raised bog units, while loss of the lagg fen is also a common feature (e.g. Joosten, Tanneberger & Moen, 2017). Sharp rectilinear shapes now often characterise the borders of surviving raised bogs (see Figure 1).

Such boundaries are common to all the surviving raised bogs in Cumbria. This has profound implications for the hydrology of these sites and means that they cannot fully meet the criterion of 'favourable condition' as defined within the EU Habitats Directive – namely that *"all structure and function necessary for the long-term maintenance of the interest are in place and likely to remain in place for the foreseeable future"*. The issue is perhaps best understood using the analogy of a lake. Abstraction of water from just one part of a lake will ultimately have an impact on the entire lake level. So it is with a raised bog system which is, after all, virtually an upturned lake, consisting as it does of some 95-98% water and with less solids per unit volume than is contained within milk. More formally, a raised bog system can be defined as a ground-water mound (Ingram, 1982) taking the shape of a half-ellipse where the degree of curvature of the ellipse is determined by the prevailing local climate. A relatively dry continental climate will give rise to a relatively flat half-ellipse whereas in a highly oceanic climate the half-ellipse will display greater curvature.

One consequence arising from the ground-water mound (GWM) model is the fact that all raised bogs within a local climate region will tend towards the same curvature of ellipse, meaning that the diameter of a bog will tend to determine its maximum height. If the diameter of a raised bog determines the maximum height of the dome, what then are the consequences of the geometric margins possessed by the various raised bogs displayed in Figure 1, and indeed all of Cumbria's raised bogs? In all these cases the original diameter of the bog has been reduced by various forms of land-claim. Calculations based on GWM theory provide some insight into the potential consequences of such reductions in original diameter.

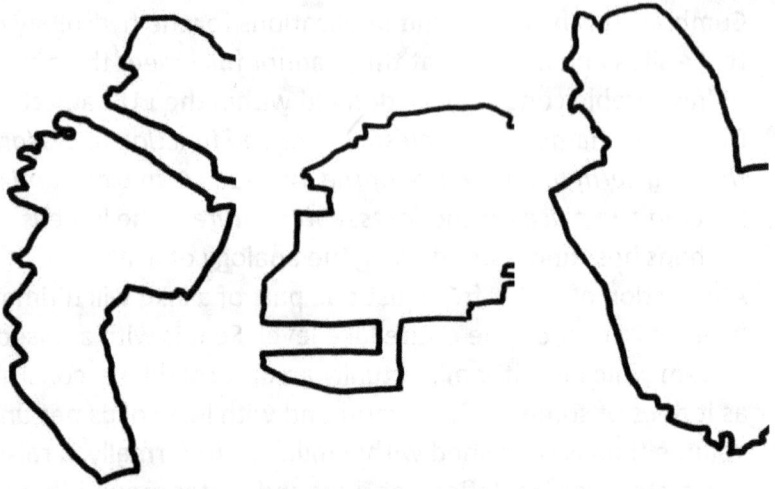

Figure 1: Outlines of three raised bog systems highlighting the un-natural geometrical shape of the raised bog margins resulting from land-claim. (Left) Duddon Mosses, Cumbria. (Centre) Burns Bog, Vancouver, Canada. (Right) Kopuatai Peat Dome, North Island, New Zealand.

Figure 2 illustrates the sequence of events that GWM theory predicts should happen if the diameter of a raised bog is reduced by some form of land claim i.e. cutting. A reduction in raised bog diameter creates a new, smaller GWM, mirroring the ellipse form of the original but with a reduced maximum height and a zone of major change arcing out from the cut face. Given that the GWM represents the bog water table, it would seem that the GWM theory predicts major falls in the bog water table across substantial parts of the site in order to regain hydrological stability. This is not, however, what is observed. There is instead a long history of hydrological literature pointing to the very limited effect of drainage associated with peatland drains and cut faces, a response attributed to the extremely low hydraulic conductivity of most raised bog peat. Drainage effects are typically described as only being significant over distances of 5 to 10 metres and rarely more than 1 metre into the peat. Setting aside for the moment the question of what might be considered a 'significant effect', such limited distances are not an indication that GWM theory is incorrect, merely that the mechanisms by which it operates are not those that are widely assumed to be the case. These are namely a measurable, substantial descent of the water-table into the peat across much of the bog.

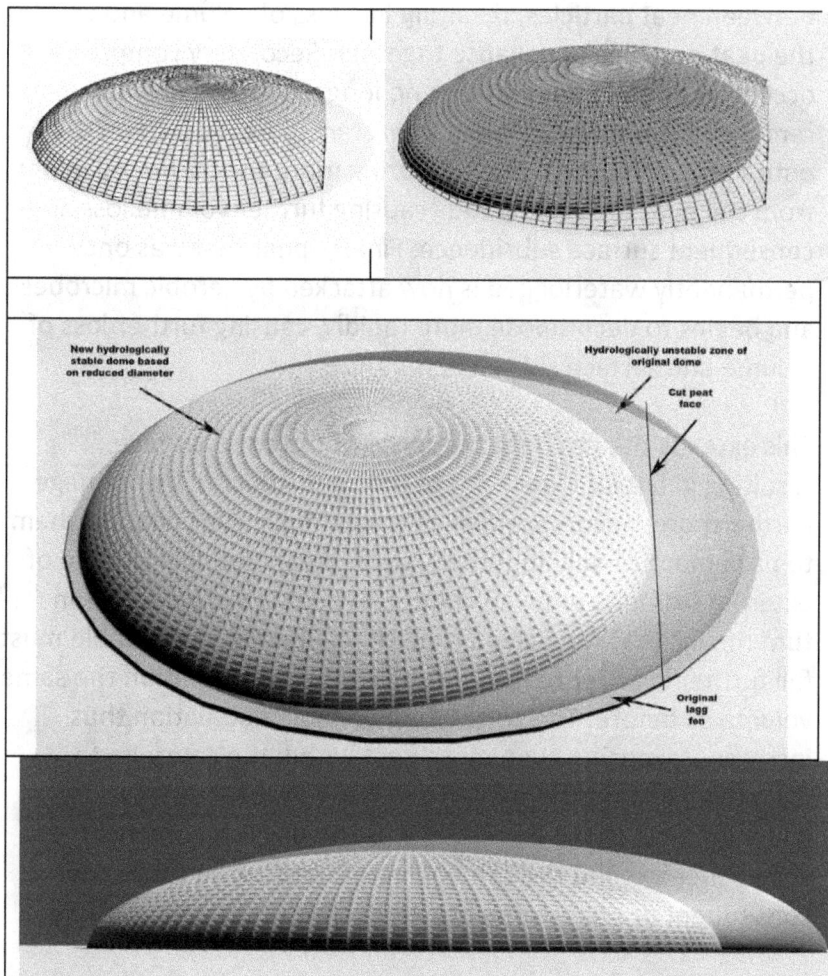

Figure 2: Raised bog half-ellipse responding to a reduction in diameter based on ground water mound (GWM) theory. (Top left) Raised bog dome with cut face, smooth grey shading indicating the original extent of the dome. (Top right) Original raised bog dome with smaller GWM formed within; based on the reduced diameter of the bog. (Middle) Features associated with a reduced GWM. (Bottom) Original and reduced GWMs illustrating the extent of impact resulting from reduction in height of the smaller GWM compared with that of the original dome.

Three processes serve to cause the peat surface to fall within a zone along the cut peat face. These processes are 1) primary consolidation, 2) secondary compression and 3) oxidative loss (Hobbs, 1986; Lindsay, Birnie & Clough, 2014). Primary consolidation results from rapid loss of water from large spaces between peat particles, resulting in a loss of volume and causing the peat particles to collapse together. Secondary compression occurs when the peat particles no longer supported in the water removed by primary consolidation start to bear down on the entire peat column below, squeezing more tightly-bound water from the peat matrix and thus causing further volume loss and consequent surface subsidence. Finally, peat that was once permanently waterlogged is now attacked by aerobic microbes and begins to decompose more rapidly, causing further loss of volume and surface subsidence.

This extends the drainage influence of the cut peat face, resulting in lateral expansion of the zone affected by drainage. Furthermore, decomposition of the peat particles enables them to pack more closely together, which mean that the volume of external storage per given volume of peat is reduced. This in turn means that for a given climate regime the water-table must fall further in order to lose through evapotranspiration the same volume as before. The zone of water-table fluctuation thus increases, exposing even more peat within the catotelm to oxidative decomposition, and this happens over an ever-widening area of the bog surface as the physical effects of subsidence expand the zone of increased surface slope and consequent drainage effect towards the cut peat face.

This combination of processes means that a feedback loop develops that will continue incrementally as long as there is a drainage effect – in other words for as long as the shape of the peat is not the same as the stable GWM for the given diameter of the peat body. In this way the peat surface 'chases the water table down', never allowing the water-table to fall more than a metre or so at most into the peat, thereby giving rise to the belief that drainage has only limited effects, yet ultimately resulting in wholesale changes to most, if not all, of the raised bog system. As the shrinking bog approaches the new stable GWM profile it can be expected that the bog surface will once again revert to a wet peat-forming system from which it can once again expand, but during the extended period of shrinkage the bog surface is likely to be – indeed must be – drier than when in its original state. This drier state may still possess an apparently peat-forming community and thus be considered to be in good health, but it will be a drier community than that which originally existed – so perhaps one dominated only by hummock-forming Sphagnum species and a number of heath species such as *Calluna vulgaris* rather than a surface pattern of pools and hummocks. It will also be a net source of both water and carbon, and display a net loss of volume over time, until the new stable profile is achieved.

Returning to the plan profiles of raised bogs displayed in Figure 1, and to all the raised bogs in Cumbria, it is clear that their original diameters have been substantially altered and therefore their hydrological stability has been compromised. This explains why many of these bogs are being invaded by tree growth, requiring significant management intervention to retain or restore open bog habitat.

Such hydrological pressures are further exacerbated by yet another factor typically found in most cases where the original bog dome has been truncated. In the natural system the base of the GWM is set by the water level in the lagg fen that represents the transition between the bog water table and the water table of the surrounding mineral ground. This lagg zone is, however, usually the first zone to be drained when land claim occurs. Furthermore, the new junction between remaining raised bog and newly-claimed land (or the lowered surface resulting from peat extraction) is typically characterised by a deep drain designed to remove water from the worked ground. As a result, the new GWM is not merely smaller and lower than the original raised bog dome, its base is also lowered by the vertical difference between the water level in the original lagg fen and the water table in the new drain running along the margin of the cut face. This factor probably has more immediate impact on the rate of surface subsidence than any other because such boundary drains are maintained regularly, and every time the drain is cleaned (and usually deepened) it stimulates a further round of rapid primary consolidation in the adjacent raised bog dome.

Conventional conservation strategies for truncated raised bog systems

A full appreciation of the implications for raised bog hydrology implicit in the GWM Theory has taken many decades to filter through into conservation thinking. However, as the implications of the GWM Theory began to be more widely understood, experiments involving various differing strategies were attempted with varying degrees of success. One of the earliest approaches involved the construction of an impermeable dam against the abrupt peat face of the site. Wheeler & Shaw (1995) and Brooks & Stoneman (1997) provide a review of bunding as applied to truncated raised bogs, informed by the GWM Theory. Wheeler & Shaw (1995) indeed proposed as one option actively reshaping the entire bog remnant to fit the modified GWM shape, although such drastic intervention was at the time largely within the context of sites subject to commercial extraction of peat for horticultural purposes, a commercial activity which no longer has government policy support and for which there is a government target to phase out completely. Brooks & Stoneman (1997) provide a detailed methodology for bunding and present a number of case studies. The approach raises a number of issues, however. Firstly, and perhaps most fundamentally, it maintains the abrupt and truncated nature of the raised bog. If successful it would be reasonable to expect the bog to continue accumulating peat and thus ultimately overtop the bund, creating yet another abrupt transition to the surrounding landscape. Such a bund also formalises the truncated nature of the bog as well as formalising the excision of the ground from the bog system, whereas any longer term answer to the GWM problem is most likely to depend on more flexible and sympathetic management of the land that was formerly part of the bog system. Fixing the artificial boundary with a structure that explicitly seeks to separate the bog from the surrounding land renders any such sympathetic management of adjacent land much less likely.

Additional issues associated with fixed bund structures arise from the contrast between the bog and the dry conditions of adjacent land. A constructed bund will experience considerable water pressure as water within the bog accumulates at the bog-bund interface, particularly towards the base of the bund. As the bund is not a naturally functioning wetland but is generally constructed from dense peat, extended periods of dry weather can lead to cracking of the bund. These cracks will not completely close during wet weather because peat undergoes irreversible shrinkage when dry. Such cracks will therefore tend to perpetuate and expand over time, leading to focused flow from the peat body and thus potential formation of structures such as large macropores, internal cracks and peat piping that can act to drain the peat body. Where the base of the bund is set into a mineral sub-soil having variable porosity, which is a not-uncommon feature at the base of raised bogs, water pressure can also force a passage through the most porous sections of the mineral sub-soil and create drainage channels (Morgan-Jones, Poole & Goodall, 2005).

Re-profiling the cut margins of the truncated raised bog dome has also been attempted, as described by Wheeler & Shaw (1995). This can be approached in two ways. Firstly, the truncated dome margin itself may be re-profiled to generate a gentler gradient to the bog margin. Two points must be considered here. Firstly, short of re-profiling the entire dome to the new smaller ellipse, any localised re-profiling at the margins will remain steeper than can be supported by the smaller ellipse, in which case little has changed and losses will continue. Secondly, the peat archive is an irreplaceable feature of a peatland. Once lost by re-shaping the peat, it cannot be restored. Any decision to re-shape the truncated edge must be taken in full recognition of the consequences.

The second approach to re-profiling the truncated margin is to push peat material from the adjacent ground up against the peat face. This approach has the advantage of not disrupting any of the 'true peat' of the truncated dome and therefore avoiding concerns about loss of the irreplaceable archive, although loss of the archive now occurs within the peat deposit of the adjacent land. If such ground is already subject to agricultural use the archive may already have been disrupted, in which case such concerns are no longer relevant. Removal of some peat material from the agricultural ground will, however, inevitably lower the surface of that ground thus increasing its tendency to flooding. This is unlikely to be a desirable consequence for the landowner.

Where the land adjacent to the truncated dome has remained within conservation control, often because such land is characterised by abandoned domestic or commercial peat cuttings, a third approach has been used, drawing on the concept of the lagg fen as the hydrological base of the GWM. In the UK the first examples of this were undertaken at Cors Caron in mid-Wales (Brooks & Stoneman, 1997), although something similar had already been applied to sites in the Netherlands. The method was termed 'pressure bunding' (although this is now applied to a wider range of bunding actions) and involved creating bunds in the areas of cut-over peat adjacent to the truncated dome. The underlying principle was that this would create a semblance of a lagg fen with a higher water table than had been the case while these areas were being exploited, and this elevated water table would induce a higher bog water table in the truncated dome because the foundation for the GWM had been elevated. Lindsay (2003) explored this method in some detail by considering the relationship between the newly-created lagg fen and the GWM of the truncated bog.

Paludiculture and raised bog conservation

A radical, if so far largely theoretical, approach to the conservation of truncated raised bogs has been explored in Austria guided almost entirely by the principle of GWM Theory.

Bragg & Steiner (1995) took the current truncated dimensions of a small raised mire, Pürgschachenmoos, and calculated what extent of hemi-ellipse shape would be required to sustain the current (steadily subsiding) height of the present raised bog remnant. They used this modelled extent to identify where a new lagg fen should be created in order to provide a sustainable hydrology for the present truncated dome, but this would only be true once the intervening ground had been managed in such a way that this ground had re-developed a peat-forming vegetation and had accumulated sufficient peat for the surface of the intervening ground to have reached the level of the truncated bog surface. The management strategy was therefore to establish this new lagg fen and encourage fresh peat accumulation across the land lying between the lagg fen and the current truncated dome. While this strategy might seem somewhat far-fetched, or at least far seeing, it is possible to identify examples where this has happened through natural recovery of raised bog systems. Heathwaite Moss in the Duddon Valley is just such an example, where a remnant dome is surrounded by vigorously infilling domestic peat cuttings with a vegetation surface which has almost grown level with the slowly sinking surface of the truncated dome (Figure 3).

However, the fundamental difference between Heathwaite Moss and Pürgschachenmoos, is that the latter is surrounded by intensively farmed agricultural land that runs right to the truncated face of the remnant dome. The reason that the conservation proposals for Pürgschachenmoos remain largely theoretical at present is that the financial cost of buying out or entering into a conservation management agreement with, the farmers who work the fields that surround the moss is so enormously high that this is not currently a feasible option. This highlights one of the key fundamental problems facing conservation and restoration actions directed towards the long-term sustainability of such truncated raised bogs. All the management efforts and all the funding expended on conservation interventions will continue to be undermined by conventional agricultural activities in the adjacent land until it becomes feasible to fund cessation of such activities.

Conservation budgets will never be large enough to achieve this on the scale required for the many surviving raised bogs – all of which are truncated – in the UK (Lindsay & Immirzi, 1996) and across much of Western Europe (Joosten, Tanneberger and Moen, 2017). This fundamental challenge will only be resolved if and when an equally fundamental change occurs within conventional agricultural practices and the support systems that underpin these.

Figure 3: Heathwaite Moss in the Duddon Valley, south Cumbria. The truncated raised bog dome can be seen as the relatively smooth pale-brown shape in the centre of the site. The mottled pale yellow and dark brown ground surrounding it represents areas of old abandoned domestic peat cuttings now filled with peat-forming vegetation.

Modern conventional agriculture in Europe originated some 9,000 years ago within the Fertile Crescent that stretched from Anatolia to Egypt and was based on species typical of semi-desert conditions. These farming practices subsequently expanded steadily westwards during the Neolithic Period, thereby establishing 'dryland' farming based on semi-desert species such as wheat, barley and sheep as the farming method of choice throughout Western Europe. Some areas, such as the East Anglian Fens, retained an economy based largely on a hunter-gatherer culture within extensive wetland areas until the end of the eighteenth century, at which point these extensive tracts of wetland not just in England but throughout much of Western Europe were converted to 'dryland' farming methods through advances brought about in quick succession firstly by the Agrarian Revolution and subsequently by the Industrial Revolution (Darby, 1940; Darby, 1956; Sheail and Wells, 1983; Williams, 1990; Rotherham, 2013; Joosten, Tanneberger & Moen, 2017). This has resulted in creation of some of the finest agricultural land. For example, the East Anglian Fens now provide 50% of England's Grade 1 agricultural land (NFU, 2019). However, conversion of wetland soils to dryland farming, not just in the UK but globally, has been achieved at a significant price. Shrinkage of drained peat soils means that substantial areas of the East Anglian Fens are now below sea level and can only be maintained as conventional agricultural systems by major pumped drainage infrastructure. The cost-benefits, or even the absolute costs, of such drainage infrastructure are therefore increasingly being questioned (Environment Agency, 2020). Such doubts are raising fears about reduced agricultural production and increased flood risk (NFU, 2019). Added to the costs of drainage infrastructure are the huge losses of biodiversity resulting from conversion of wetland systems to dryland systems (IPBES, 2018; European Union, 2020). There are also, perhaps most pressingly given the current climate emergency, the massive emissions of carbon from drained peatlands. In the UK alone, agriculturally worked peat soils represent the largest single source of carbon emissions from land use (Evans *et al.* 2017a) and much the same picture is true across Western Europe (Joosten *et al.*, 2016).

These consequences have arisen because wetlands have been destroyed and converted to dryland conditions in order to enhance agricultural productivity - yet wetland systems represent some of the most productive ecosystems on Earth (Williams, 1990). Some wetland species can rival the best wheat production rates obtained from the North American prairies (see Figure 4), but until now there has been little incentive to explore or develop modern methods of production and new potential markets for wetland species because agricultural support systems have consistently favoured conventional 'dryland' crops.

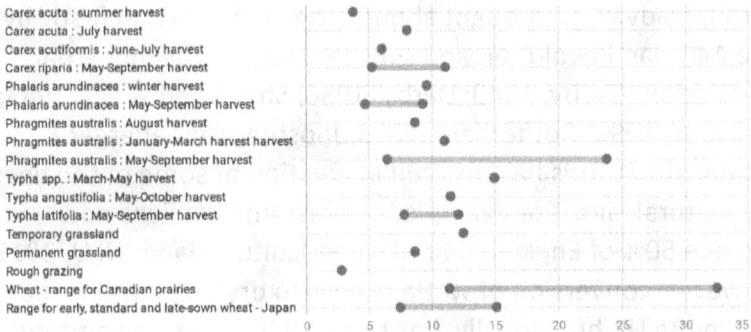

Carex acuta : summer harvest							
Carex acuta : July harvest							
Carex acutiformis : June-July harvest							
Carex riparia : May-September harvest							
Phalaris arundinacea : winter harvest							
Phalaris arundinacea : May-September harvest							
Phragmites australis : August harvest							
Phragmites australis : January-March harvest harvest							
Phragmites australis : May-September harvest							
Typha spp. : March-May harvest							
Typha angustifolia : May-October harvest							
Typha latifolia : May-September harvest							
Temporary grassland							
Permanent grassland							
Rough grazing							
Wheat - range for Canadian prairies							
Range for early, standard and late-sown wheat - Japan							
	0	5	10	15	20	25	30	35

Figure 4: Average yields for a range of wetland and agricultural crops (tonnes of dry matter per hectare per year). Sources: Huffman *et al.* (2015); Oehmke & Abel (2016); Qi *et al.* (2018); Sawada *et al.* (2019).

Nevertheless, perceptions are now changing. There is recognition that unless action is taken, agricultural productivity in areas currently rich in peat soils will inevitably fall. Furthermore the costs of flood protection are spiralling but despite huge investment the challenges of holding back flood waters are starting to prove insurmountable – the reality of which is leading both the insurance industry and official bodies responsible for water management to question conventional wisdom and 'business as usual' (Gremli *et al.*, 2013; Zurich Flood Resilience Alliance, 2019; Environment Agency, 2020). Finally, and the driving force responsible for focusing much inter-governmental thinking about the issue, is recognition that something must be done to halt the enormous carbon emissions associated with current agricultural use of peat soils (Evans *et al.*, 2017a; Evans *et al.*, 2017b*).*

These various drivers have led to a re-examination of wetland species that either in the past were proved to be economically useful or which now offer potential to meet current market demands. The term 'paludiculture' has been coined to describe the farming of wetland species under wetland conditions and is now the subject of a great many pilot studies and field trials in various parts of the world (Joosten *et al.*, 2012; Biancalani & Avagyan, 2014). In some cases the trials are focusing on the potential for commercial expansion of established wetland crops such as reed (*Phragmites australis*) while in other examples entirely new crop species and entirely novel uses are being explored. It is important to recognise that paludiculture is not nature conservation. It is an activity focused on generating a commercially viable crop through agricultural practices. The fact that nature may benefit in some way from paludiculture is an unintended side-effect and one example of the potential co-benefits arising from its adoption, along with, for example, reduced flood-risk, greater flood storage, reduced ground subsidence and halting carbon emissions.

That such a new approach to agriculture is being taken seriously is evident in recommendations from the UK Committee for Climate Change, the UK Government's 25-Year Environment Plan, the guidance being provided by the UN FAO about the subject, and an increasing number of high-level statements and actions concerning the topic. The potential market for paludiculture products is also increasingly being recognised as both potentially lucrative and very large (Wichtmann, Schroder and Joosten, 2016). One of the largest centres of activity currently investigating the processes and products of paludiculture is to be found in Germany (Gaudig, 2014; Wichtmann, Schroder and Joosten, 2016), but a growing number of experimental and field-scale trials have now been established in the UK (Mulholland *et al.*, 2020).

These and other initiatives around the world offer a valuable way forward for raised bog conservation in promoting the production of commercially viable wetland crops from wet peat soils instead of seeking to drain them for conventional agriculture. Although carbon emissions are greatest from peat soils that are Grade 1 agricultural land subject to arable agriculture, debate about the cost-benefits of flood risk, flood storage and sea defences are focused more on lower grades of agricultural land. Little, if any, land adjacent to truncated raised bogs in Cumbria is classed as Grade 1 agricultural land but much of it is at significant flood risk and also has considerable potential to offer in terms of flood storage. Furthermore, while carbon emissions from such areas are nothing like as high as those from arable peat soils in the Cambridgeshire Fens they are still significant. In addressing these issues by adopting the higher water tables associated with paludiculture, such actions would also have the added benefit of providing a new type of lagg fen with a high water table at the margins of truncated raised bogs, thus addressing one of the most fundamental challenges currently affecting these sites. Adoption of paludiculture in land adjacent to such truncated raised bogs would thus mean a series of multiple wins:

- The hydrological condition of the raised bog would be improved;
- This enables the input of conservation funding and resourcing to be more effective;
- The farmer would no longer need to expend resources on continually draining the adjoining land;
- Potential exists for higher-value crops to be grown on this land than is currently possible under conventional agriculture;
- Land subsidence can be halted across the affected fields;
- Perception of flood risk for the affected fields is fundamentally altered;
- Local flood storage can be increased;
- Local/regional flood risk can be reduced;
- National carbon accounting benefits from reduced carbon emissions;
- Pressure on the global atmosphere from carbon emissions is reduced, or even reversed into carbon storage.

The potential offered by paludiculture is therefore considerable but potential must not be allowed to run ahead of pragmatism. Conventional farming has had 9,000 years to develop agricultural systems that work to greater or lesser degrees. Paludiculture has had a mere 10 to 15 years devoted to developing an entirely new agricultural system. There is much still to learn about potential crops, their husbandry and harvesting, the machinery required to achieve this cost-effectively, the products and markets that might emerge from this new approach to land management, as well as consideration of the implications for local and even regional infrastructure that might arise from widespread adoption of paludiculture. In some cases at least, even such unpredictable ventures into the unknown are no worse than the pressures that are increasingly bearing down on conventional farming systems with no realistic prospect of improvement, so despite the uncertainties it may be that paludiculture at least offers some hope for the future. Finally, and potentially most important of all, is the stimulus that could be provided through government farm support systems to encourage adoption and development of this entirely new farming system of the future – an entirely new farming system of the future whose ancient origins lie in the lake villages of the pre-Neolithic landscape. Perhaps the time has finally arrived for mainstream adoption of paludiculture and thus also, as a happy co-benefit, for a fundamental improvement in the condition of our lowland peatlands.

References

Biancalani, R., & Avagyan, A. (eds) (2014) *Towards climate-responsible peatlands management. Mitigation of Climate Change in Agriculture Series 9.* Food and Agriculture Organization of the United Nations, Rome.
Bragg, O.M., Lindsay, R., Robertson, H., and others (1984) *An Historical Survey of Lowland Raised Mires, Great Britain.* Joint Nature Conservation Committee, Peterborough.

Bragg, O.M., & Steiner, G.M. (1995). Applying groundwater mound theory to bog management on Pürgschachenmoos in Austria. In: Moen, A. (ed.) Regional variation and conservation of mire ecosystems. *Gunneria,* **70,** 83-96.

Brooks, S. & Stoneman, R.E. (1997) *Conserving bogs: the management handbook*. Stationery Office, Edinurgh.

Burton, R.G.O. & Hodgson, J.M. (eds) (1987) *Lowland Peat in England and Wales. Soil Survey Special Surveys No. 15*. Soil Survey of England and Wales, Cranfield.

Darby, H.C. (1940) *The Medieval Fenland*. Cambridge University Press, Cambridge.

Darby, H.C. (1956) *The draining of the Fens (2nd ed.)*. Cambridge University Press, Cambridge.

Environment Agency (2020) *National Flood and Coastal Erosion Risk Management Strategy for England*. Environment Agency, Rotherham.

European Union (2020) *Conclusions on Biodiversity - the need for urgent action – Approval*. Council of the European Union, Brussels.

Evans, C., Artz, R., Moxley, J., Smyth, M-A., Taylor, E., Archer, N., Burden, A., Williamson, J., Donnelly, D., Thomson, A., Buys, G., Malcolm, H., Wilson, D., Renou-Wilson, F. & Potts J. (2017a) *Implementation of an emission inventory for UK peatlands. Report to the Department for Business, Energy and Industrial Strategy*. UK Centre for Ecology and Hydrology, Bangor.

Evans, C., Morrison, R., Burden, A., Williamson, J., Baird, A., Brown, E., Callaghan, N., Chapman, P., Cumming, A., Dean, H., Dixon, S., Dooling, G., Evans, J., Gauci, V., Grayson, R., Haddaway, N., He, Y., Heppell, K., Holden, J., Hughes, S., Kaduk, J., Jones, D., Matthews, R., Menichino, N., Misselbrook, T., Page, S., Pan, G., Peacock, M., Rayment, M., Ridley, L., Robinson, I., Rylett, D., Scowen, M., Stanley, K., & Worrall, F. (2017b) *Lowland peatland systems in England and Wales – evaluating greenhouse gas fluxes and carbon balances. Project code: Defra SP1210*. UK Centre for Ecology and Hydrology, Bangor.

Gaudig, G., Fengler, F., Krebs, M., Prager, A., Schulz, J., Wichmann, S., & Joosten, H. (2014) Sphagnum farming in Germany – a review of progress. *Mires & Peat*, **13**, Article 08, 1-11.

Gremli, R., Keller, B., Sepp, T., & Szönyi, M. (2013) European floods: using lessons learned to reduce risks. Zurich Insurance Group, Zurich.

Hobbs, N.B. (1986) Mire morphology and the properties and behaviour of some British and foreign peats. *Quarterly Journal of Engineering Geology*, **19**, 7-80.

Huffman, T., Qian, B., DeJong, R., Liu, J., Wang, H., McConkey, B.G., Brierley, T. & Yang, J. (2015) Upscaling modelled crop yields to regional scale: A case study using DSSAT for spring wheat on the Canadian Prairies. *Canadian Journal of Soil Science*, **95**, 49-61.

Ingram, H.A.P. (1982) Size and shape in raised mire ecosystems: a geophysical model. *Nature*, **297**, 300-303.

Joosten, H., Tapio-Biström, M.-L., & Tol, S. (eds) (2012) *Peatlands - guidance for climate change mitigation through conservation, rehabilitation and sustainable use, 2nd edition. Mitigation of Climate Change in Agriculture Series 5.* Food and Agriculture Organization of the United Nations, Rome and Wetlands International, Wageningen.

Joosten, H., Sirin, A., Couwenberg, J., Laine, J. & Smith, P. (2016) The role of peatlands in climate regulation. In: Bonn, A., Allott, T., Evans, M., Joosten, H., & Stoneman, R. (eds.) *Peatland Restoration and Ecosystem Services: Ecological Reviews.* Cambridge University Press, Cambridge, 82-95.

Joosten, H., Tanneberger, F., & Moen, A. (eds) (2017) *Mires and peatlands of Europe – status, distribution and conservation.* Schweizerbart Science Publishers, Stuttgart.

Lindsay, R. (2003) Peat forming process and restoration management. In: Meade, R. (ed.) *Proceedings of the Risley Moss Bog Restoration Workshop 26-27 February 2003*. English Nature, Peterborough, 23-38.

Lindsay, R., & Immirzi, P. (1996) *An inventory of lowland raised bogs in Great Britain*. Scottish Natural Heritage, Perth.

Lindsay, R., Birnie, R. & Clough, J. (2014) *Peat Bog Ecosystems: Impacts of Artificial Drainage on Peatlands*. International Union for the Conservation of Nature, Edinburgh.

McTeague, E., & Watson, K. (1991) *A peatland Survey of Mid-Strathclyde, Scotland (1989)*. Nature Conservancy Council, Peterborough.

Montanarella, L., Scholes, R., & Brainich, A. (eds) (2018) *The IPBES assessment report on land degradation and restoration*. Secretariat of the Intergovernmental Science-Policy Platform on Biodiversity and Ecosystem Services, Bonn, Germany.

Morgan-Jones, W., Poole, J., & Goodall, R. (2005) *Characterisation of Hydrological Protection Zones at the Margins of Designated Lowland Raised Peat Bog Sites. JNCC Report No. 365*. Joint Nature Conservation Committee, Peterborough.

Mulholland, B., Abdel-Aziz, I., Lindsay, R., McNamara, N., Keith, A., Page, S., Clough, J., Freeman, B., & Evans, C. (2020). *Literature Review: Defra project SP1218: An assessment of the potential for paludiculture in England and Wales.* UK Centre for Ecology & Hydrology, Bangor.

NFU (2019) *Delivering for Britain: Food and Farming in the Fens.* NFU, East Anglia.

Oehmke, C. & Abel, S. (2016) Promising plants for paludiculture. In: Wichtmann, W., Schröder, C., & Joosten, H. (eds.) *Paludiculture – productive use of wet peatlands. Climate*

protection – biodiversity – regional economic benefits.
Schweitzerbart Science Publishers, Stuttgart, 22-38.

Qi, A., Holland, R.A., Taylor, G., & Richter, G.M. (2018) Grassland futures in Great Britain – Productivity assessment and scenarios for land use change opportunities. *Science of the Total Environment,* **634**, 1108-1118.

Rotherham, I.D. (2013) *The Lost fens: England's greatest ecological disaster.* The History Press, Stroud.

Sawada, H., Matsuyama, H., Matsunaka, H., Fujita, M., Okamura, N., Seki, M., Kojima, H., Kiribuchi-Otobe, C., Takayama, T., Oda, S., Nakamura, K., Sakai, T., Matsuzaki, M., & Kato, K. (2019) Evaluation of dry matter production and yield in early-sown wheat using near-isogenic lines for the vernalization locus Vrn-D1. *Plant Production Science,* **22** (2), 275-284.

Sheail, J. & Wells, T.C.E. (1983) The fenlands of Huntingdonshire, England. A case study in catastrophic change. In: Gore, A.J.P. (ed.) *Mires: swamp, bog, fen and moor. Ecosystems of the world 4B.* Elsevier, Amsterdam, 375-393.

Wheeler, B.D., & Shaw, S.C. (1995) *Restoration of damaged peatlands – with particular reference to lowland raised bogs affected by peat extraction.* HMSO, London.

Wichtmann, W., Schroder, C., & Joosten, H. (eds) (2016) *Paludiculture – productive use of wet peatlands: Climate protection – biodiversity – regional economic benefits.* Schweitzerbart Science Publishers, Stuttgart.

Williams, M. (1990) Understanding Wetlands. In: Williams, M. (ed.) *Wetlands – A Threatened Landscape.* Basil Blackwell Ltd, Oxford, 1-41.

Zurich Flood Resilience Alliance (2019) *The Flood Resilience Measurement for Communities (FRMC).* Zurich Flood Resilience Alliance, Zurich.

Conference field visit – photograph © Solway Connections Guided Heritage Tours

Chapter 4. Angerton Moss: a peat resource from the thirteenth to the twentieth century

William D. Shannon

Summary

Angerton Moss, now in Cumbria, was within Lancashire until 1974. A small amount of living raised mire or 'moss' remains, within the Duddon Mosses National Nature Reserve. Place-name evidence suggests early-medieval use of the area for hunting and summer grazing, but the first mention of the Moss occurs in the mid-thirteenth century, at which date it was clearly being exploited primarily as a fuel resource. At the end of that century, the Moss was granted to Furness Abbey, and it remained in monastic ownership until the Dissolution, with the abbey's tenants throughout Low Furness having rights to cut peat on the Moss. Shortly before the Dissolution, the first enclosures took place, the start of a process which continued over the following centuries, first under the Crown, and later under the succeeding owners. Evidence for the landscape of the moss in the Elizabethan age comes from a fine manuscript dispute map, now in the National Archives. At that date, a large acreage of turbary remained, surrounded by extensive areas of rough pasture and salt marsh, as well as enclosures of improved land. That picture remained largely unchanged for 200 years, but the eighteenth and early nineteenth centuries saw increased reclamation and occupation: although peat continued to be dug for fuel, as can be seen from a sale catalogue of 1902, which allows us to understand the landscape at that date. Oral accounts tell us that the locals from all the farms within Angerton continued to cut peat for their fires up to the last war – but there was also some commercial peat production going on into the first half of the twentieth century, immediately to the north of Angerton Moss, where a Moss Litter Works had been established, with a tramway taking the peat, used for horse bedding, down to the coast.

Keywords: *Lancashire, Furness, Abbey, Dissolution, moss, turbary, enclosure, fuel, dispute map, moss litter.*

Introduction

The present study arose from work carried out in association with the Regional Heritage Centre of Lancaster University and the Cumbria County History Trust, as part of the *Victoria County History of Cumbria* project, which aims to write a brief history of every one of the 348 places (villages, townships or parishes) which make up the present day county of Cumbria, following a standard format covering population, economy, landownership, places of worship, and institutions. Around a dozen draft histories have been written so far, and posted to the Cumbria County History Society website, including one for Angerton (CCHT, 2015). This supersedes the very brief entry on Angerton which had appeared in 1914 in Volume 8 of the original *Victoria County History of Lancashire* (Farrer & Brownbill, 1914, 408-9).

From the thirteenth century, when it was acquired by Furness Abbey, until 1857, Angerton Moss was an extra-parochial district. From 1857 to 1976, it was a Civil Parish; until 1974 as part of Lancashire. Since 1976 it has been part of Duddon parish, within South Lakeland District, in the county of Cumbria. It is bounded by the historic parishes of Broughton in Furness to the north, and Kirkby Ireleth, or Kirkby in Furness, to the east and south. To the west, it is bounded by the estuary of the River Duddon which formed the historic boundary between Lancashire and Cumberland.

Originally given over almost entirely to raised mire and salt marsh, only a small extent of living moss now remains, distinguished by sphagnum and bog cotton, with the drier parts covered with heather and grasses, now lying within the Duddon Mosses National Nature Reserve which straddles the former border between Angerton and Broughton. The Soil Survey (see Soilscapes website) shows that extensive areas of Raised Peat Bog Soils remain in the area covered by the Nature Reserve - naturally wet, and with very low fertility. To the south, though, the mosses have disappeared, and the former turbary and open seasonal pasture have been replaced by enclosed and settled farm land, largely down to grass, on soil which is still peaty, but fertile enough when drained. Only the place-name Moss Houses remains as a reminder of the former landscape.

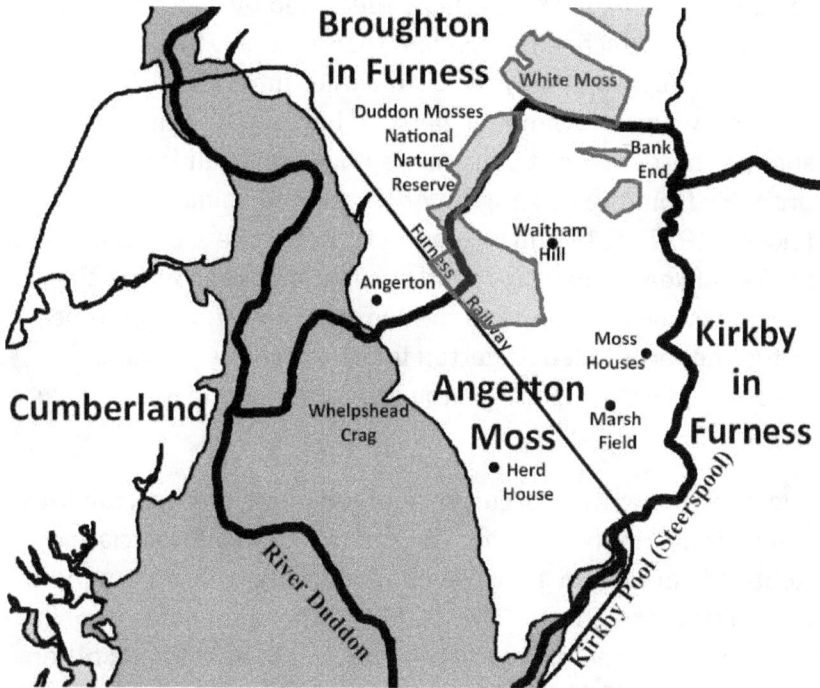

Figure 1. Angerton Moss, showing places named in the text. Five scattered parcels making up Duddon Mosses National Nature Reserve are also shown

Earliest landscape and economy

The earliest landscape can be reconstructed from the place-names shown on the above map. The first of two Old Norse names recalls a time when the main economic value of the area was hunting: Waitham, *veidi-holmr*, 'island from where hunting is carried out' (Ekwall, 1922, 189,198). The use of the word 'island' here (and the later addition of the word 'Hill') suggests a slightly higher, and hence drier, location used as a hunting station, at the time surrounded on all sides by wetlands which have since been drained and tamed. The other name is equally evocative, Whelpshead Crag, from *hvelpr* (ON), a whelp or youngster, and *saetr* (ON) or *set, sat,* (ME) a shieling or summer settlement (Ekwall, 1922, 222). This might suggest a slightly later time when there was temporary seasonal occupation on this elevated point above the estuary, perhaps surrounded by sea at the highest tides, but from where the young people could spend the summer watching over the herds grazing on the salt marsh. The same idea is perhaps suggested by another name, Herd House, which, although first recorded only in the nineteenth century, may recall an earlier time, medieval or early modern, when a communal 'herd' or herdsman spent the summer here. However, the name Angerton itself is most probably from the Old English *anger tun*, meaning pasture farm (Ekwall, 1922, 221), and suggests at some stage around a thousand years ago year-round occupation began at the site which has given its name to the Moss. However, it should be noted the farm called Angerton is not within the 'extra-parochial place' called Angerton Moss (see boundaries in Figure 1, above).

The very first written occurrence of the name of Angerton Moss can be found in the *Coucher Book* of the nearby Cistercian Abbey of Furness, in an agreement which dates from around the mid-thirteenth century (TNA DL 42/3). The agreement is mainly concerned with an exchange of lands, but in it, amongst other things, John of Kirkby Ireleth granted to Ralph Fitzalan the fishery of Steerpool (now Kirkby Pool) – while in return Ralph granted to John 40 loads of turf a year from Angerton Moss, in perpetuity, to be burned in John's manor house at Kirkby. The Latin word used here for 'cart loads of turf', '*charecta turbarum*', is of some interest, as it derives from 'chariot', which is itself not Latin, but borrowed from Gaulish: although unlikely, this might conceivably suggest the survival of a local word related to the Old Welsh word for a cart or waggon, *carr*. In this context, the clerk was probably thinking of a two-wheeled cart pulled by a pony (donkeys were not common in England at this date). Its importance lies in implying the existence as early as this of a well-regulated system of peat-digging, and transport of the peat off the Moss. The reason the monks had recorded this transaction was because they were, later in the same century, to be given the moss by one Thomas Skillar, a burgess of the nearby town of Dalton - but first in 1292 they recorded the purchase of the Moss by Thomas from Richard FitzSimon of

Broughton (TNA DL 42/3*).* Up to this date, it would appear that it had been an uninhabited parcel of waste, operated in part as a turbary and in part as common pasture, belonging mainly or entirely to Broughton manor within the barony of Ulverston. In this first description of the Moss it is said to comprise 200 acres of turbary and 100 acres of pasture, with the likelihood that these would have been local customary acres, about twice the size of statute ones.

Shortly afterwards, in 1299, Thomas Skillar granted to the monks of Furness Abbey. *'...all lands, woods, pasture and mosses* [mussam] *... in the place which is called <u>Angartunmos</u> in the town of Ulverston'.* The grant describes the boundaries in some detail, and those boundaries continued to be the boundaries of Angerton Moss for as long as it remained an 'extra-parochial place' lying between Kirkby and Broughton, although never subsequently belonging to either of them. Both manors continued to claim rights within it, however; and not just that right for the lord of Kirkby to have forty cart-loads of turf a year out of the Moss, but numerous other claims from Broughton, and other places throughout the whole of the Furness peninsula, leading to court cases in the Duchy of Lancaster's Chancery Court at Westminster, which are now in the National Archives, and which allow us to reconstruct the history of the Moss both under the monks, and after the Dissolution.

Monastic and early-modern Angerton

From the start, the abbey's sole ownership did not go unchallenged. In 1424, the lord of Kirkby in Furness claimed it was he rather than the monks who owned the freehold of the Moss, and that he had the right to cut 270 cart loads of turf a year (TNA DL 25/398). Assuming a couple of loads per household per year, that would have been enough for well over 100 families. Probably historically the tenants of Kirkby really did have peat-cutting rights here, though probably not grazing rights on the moss. Either way, they lost their case. However, in part compensation the court awarded the lord of Kirkby and his tenants the right in perpetuity to cut 80 cart loads of turf a year within the abbey's moss – enough for say 40 families.

A hundred years later, on the eve of the Reformation, we have an abbey rental (TNA SC 12/9/73). From this we learn that Angerton Moss contributed a modest twenty-eight shillings and four pence rent a year to the abbey, although out of that they had to pay five shillings to Lord Derby, who was lord of the manor of Broughton - and who would appear, like the lord of Kirkby, to have continued to have had some claim on the Moss. This rental probably included something from a recent enclosure, which had led to the first permanent settlement in Angerton, at Moss Houses, which will be discussed shortly. However, the rental also tells us a great deal about how the tenants of the abbey throughout Furness were also using the moss. Thus for example four hamlets on Walney Island, ten or so miles away – Southend, Biggar, Northscales and Northend - each had the right to cut turf, but in exchange had to supply the abbey with twenty cart loads of turf a year – eighty in all – with a monetary value of £2. As one load was worth 6d, which was approximately a labourer's wages for two days at the time, this equates to six man-month's work keeping the warming room and the kitchen at Furness supplied with free fuel. It is also worth noting that this fuel was worth far more than the cash rental income from the Moss,

Then in 1537, the abbey was dissolved, its property being confiscated by the Crown and absorbed into the Duchy of Lancaster (TNA E322/91). Angerton Moss was then leased out as a separate entity by the Duchy to Thomas Preston, who had acquired the abbey site – and to Roger Kirkby, lord of Kirkby in Furness. This led immediately to disputes, as former abbey tenants all over the Furness peninsula continued to claim their ancient rights on the Moss, which the new lords challenged (TNA DL3/48/R5). Those claims were upheld in 1545, when the former tenants' rights to dig turf for their own use were confirmed, although their claim to have rights of pasture as well as turbary were rejected – they could only pasture their animals during the couple of weeks they were camping on the Moss digging turf – or at those times when they got cut off by the tide whilst trying to lead their turves home, which involved fording Steerspool (Kirkby Pool), which had no bridge.

There were no inhabitants living permanently in any part of Angerton Moss until a couple of cottages were erected at Moss Houses in the early sixteenth century, probably as part of an enclosure campaign carried out by the monks, or with their encouragement. Then some fifty years later we hear of a second enclosure episode, with a cottage, possibly also at Moss Houses (TNA DL 1/137/R12). In 1586, some 124 customary acres (= c.200 statute acres) of pasture and moss in Angerton Moss, with a cottage, had been let by the Duchy to John Richardson and Leonard Rawlinson of Furness for thirty-one years, (TNA DL 1/137/R12). Having secured their lease, the two new lessees promptly started litigation, challenging the claimed rights within the moss of the two adjacent lords, Preston and Kirkby; and a commission was set up by the Duchy to sort it out. But during the court hearing, a whole new issue was revealed – the Duddon had apparently recently changed its course to the west, thereby creating 150 acres of new land – which Lord Derby (lord of Broughton, to the north of the Moss) claimed as his, although by rights it probably belonged to the Crown. In order to ensure that the Queen was not losing out, the Duchy Court ordered a map to be made, showing the extent of the Moss, and of the new marsh lands (TNA MPC 1/34). The court then decided on the strength of this evidence that it was indeed the Queen who owned this new land, not Lord Derby – but the map tells us much more than that, allowing the landscape and economy of Angerton Moss towards the end of the sixteenth century to be accurately reconstructed.

The map was made using ink and colour wash on parchment, and it is typical of its period in focussing primarily upon what is in dispute – that is, upon the extra-parochial area of Angerton Moss, together with those parts of Broughton, to the north, which were also mossland, and over which Lord Derby was lord. It was a scale map, and was signed off for accuracy by the commissioners. It was also signed by the surveyor, Edmund

Moore, who we know also made a number of other high-quality maps for the Duchy, in and around Lancashire (Shannon, 2010, 2012).

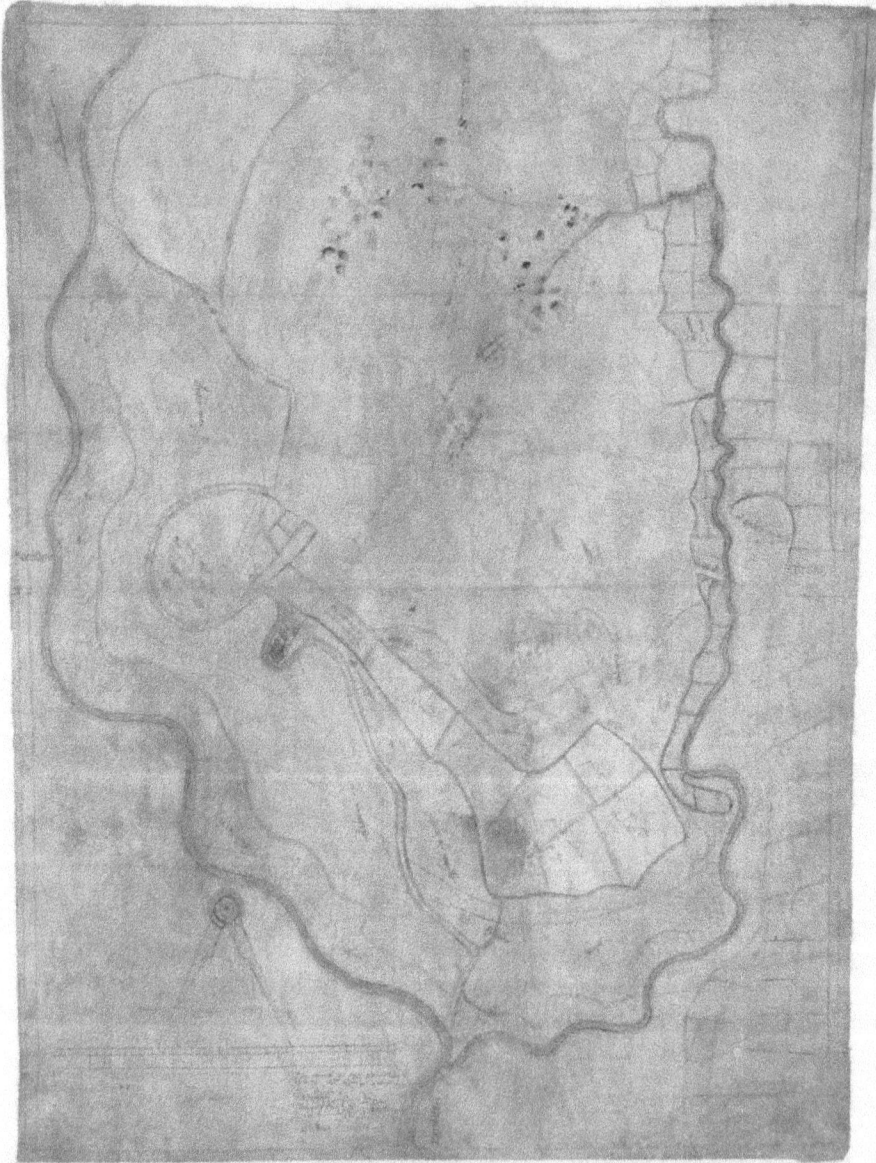

Figure 2. Edmond Moore's Map of Angerton Moss, 1587, showing the old and new courses of the River Duddon, the new salt marsh, and recent enclosures. (TNA MPC 1/34: Reproduced by permission of the National Archives)

The map clearly shows the problem: the River Duddon has shifted from its old easterly course close up by Whelpshead to a new course nearer the Cumberland side of the estuary, and in so doing has exposed a new salt marsh, to the south of Angerton Farm, and west of the old course of the river. However, Moore also draws our attention to two areas of enclosed fields, standing out to east and west of the unenclosed moss. To the west, between the moss and the old river course, are shown the '*Several* [enclosed] *Lands of the Queen*'. These are the recent enclosures which had given rise to the court case: but these are distinguished from the old, monastic, enclosures to the east of the moss, stretched out along the line of the Steerepool, one small field deep, and including the settlement of Moss Houses. Away from the enclosures, the interior of the moss is marked with a number of red blotches, which represent the peat diggings to the north of Waitham Crag, differentiating between the mosses belonging to Lord Derby as lord of Broughton – and those of the Queen as lord of Angerton. South of Waitham, though, there is no dispute about ownership of the Moss, as all here belonged to the Queen – and it was on the fringes of that moss that the new enclosure and reclamation had been taking place.

We can thus use this map to build up a clear picture of the landscape in the 1580s. A very large acreage of living moss and turbary remained, split between Angerton and Broughton. Surrounding the moss were extensive areas of rough pasture, some of which was salt marsh, covered at the highest tides, and including the new marsh. Then alongside the estuary, there were extensive tidal sands, covered by sea twice a day. In addition, there were the two areas of enclosure, one recent, and one dating back to monastic times, both having been reclaimed from the mossland, possibly following exhaustion of the peat, and probably involving both drainage and other techniques such as paring and burning.

What we are seeing in early-modern Angerton is fairly typical of the progress of the reclamation of a moss (Shannon, 2015). As a generalisation, mosses lay between rather than within townships, and served as a common resource. In the centre was what was often referred to throughout Lancashire as the *White Moss*, the living raised mire, impossible to drain without wind, steam or electric pumps, and hence of little value then for either grazing or peat. This was sometimes surrounded by a partially drained area of *Grey Moss*, poor quality peat, subject to flooding. Further out was the *Black Moss*, the best quality peat, artificially drained and laid out into moss rooms, leased to the tenants of lords of the township(s) – but the peat in this part was not necessarily very deep. Over time, as it got stripped away, what was known as the *Following Land* emerged, to be used as pasture, or improved and enclosed for arable. Over time, each zone tends to push in upon the next, until nothing remains of the original moss.

From the evidence of the court cases and the *Coucher Book*, we know that the monks of Furness had eighty loads a year for their warming room and kitchen at the abbey. As a guess, we could say that this was perhaps a tithe (tenth) of what the tenants were taking. If we surmise that the tenants from Broughton and Kirkby were also taking eighty loads a year each, then that adds to about 1,000 loads in total, say 1,000 cubic yards, each year for 500 years (not counting all the preceding years), which equates to 100 acres of peat cleared to a yard deep – and dug predominantly from the mossland edges where the peat would sometimes have only been a yard deep to begin with. Whilst these figures cannot pretend to be accurate, they can give some impression of an order of magnitude, and show how a relatively small amount of turf per household per year (two or three loads each) can, over time, clear and convert a substantial acreage of moss rooms into farm land.

Eighteenth- and nineteenth-century reclamation and enclosure

By the early seventeenth century, Angerton Moss had become divided into three parcels, as King James I sold off his possessions. To the west, the land which the Prestons had leased from the Crown since the Dissolution was bought by them in 1608, while the central and eastern part was bought from the Crown by Lord Derby. Both of these parcels contained a mixture of turbary and enclosures; but the third parcel, to the north, which comprised only the poorest quality raised mire, which no-one wanted, remained in Crown hands.

This change of ownership once again sparked a legal dispute, as tenants throughout Furness continued to claim their turbary rights (TNA DL 4/83/50). The men of Kirkby had their moss rooms set out, but were complaining that other men from throughout Furness were encroaching upon their turf – and the officials known as the four sworn men of Kirkby could not do anything about it. A witness called Robert Gibson, from near Barrow, argued that he, like all his majesty's tenants, still had the right to cut turf in Angerton Moss all the way up to Waitham, in their known moss rooms. Another witness, Christopher Brownrigge, a tenant from the Kirkby side of the moss, told how the moss rooms he claimed had been set out with fixed boundaries some twenty years previously, which had apparently involved some cost in *'dressing and befitting'* the Moss so that the peat could be extracted – which would certainly have involved drainage to some extent – and which was, of course, the first step on the road towards reclamation; although there is not much sign of further enclosure during the rest of the seventeenth century.

As we move into the eighteenth century, the landholdings change again, with first the Lowthers, then the Cavendishes, later Earls of Burlington, taking over from the Prestons, while Lord Derby had sold out as a result of the Civil War, his land eventually coming to be held by the Towers family of Duddon Hall. Meanwhile General Monk, who had done very well out of the Restoration, was granted the Crown holding in the north of Angerton, which subsequently descended, like much of Furness, to the Dukes of Buccleuch. We know that the Holker estate of the Lowthers continued to exploit the moss for grazing, as can be seen from a rental of 1726 (LA DDCA/4/2), at which date they were getting around eleven shillings a year from each of some twenty-four tenants leasing grazing in Angerton: but they were also getting turbary rents too, with some seventy-seven households still leasing moss rooms, and paying on average about nine pence each *per annum*. This of course only refers to that part of Angerton that the Holker estate held – roughly a third.

The new owners do not seem to have encouraged enclosure, and William Yates' map (1786), two hundred years after Moore, seems to show relatively little change to the moss since Edmund Moore's time. The area is still dominated by open, unenclosed mossland, surrounded by the old monastic/Elizabethan enclosed farm land. The only residential occupation would appear still to be the three houses at Moss Houses. However, the Tithe Map, copied from a survey of 1805, shows considerable progress in the interim, with new enclosures between Moss Houses and Waitham Hill, which must have been first occupied at this time (TNA IR 30/18/14). These new enclosures can be seen on Greenwood's map (1818), which shows clearly the encroachment that has proceeded since Yates' time onto the former Moss west of Moss Houses to beyond the unnamed Waitham Hill. Hennet's map (1830), a dozen years later, does not show the Moss as such – but he names Waitham Hill for the first time on a map– and also shows a new property at Herd House, on a site which, as suggested earlier, may have long been used as a temporary summer shelter for a common herdsman.

The Tithe Commissioners' report of 1839 (TNA IR 18/3895) gives us the first independent account of the quality of the land:

'The subsoil is chiefly peat, some Rag or Trap rock, and some gravel [this is in the vicinity of Waitham Hill]. *The land is poor in quality and above half of it peat. The part of it, nevertheless, which is capable of cultivation, is in general well farmed. High Farming would be thrown away upon it.'*

In other words, the Commissioners believed that there was no point in using artificial fertilisers or other soil improvement here, as the land would not make a return on that investment. We go on to learn from the Tithe Commissioners that by this date there were 469 acres of arable and 303 of meadow and pasture, the latter including *'A Large tract of mossland partially pastured'*. But there are clearly some definition problems here. We are told that the 469 acres, although designated as arable, in fact were what the locals called *'Hard Land'*, to distinguish it from the soft mossland – and that the so-called 'arable' in fact was used as well for meadow or pasture as for the plough. There is no mention here of turf (as peat was not titheable), but there was no doubt turf cutting still continuing on the soft land, as we shall see shortly.

Then, a dozen or so years later still, the First Edition Six Inch Ordnance Survey (1850) shows in detail the full extent of enclosure to date, including some further recent enclosure on Herd House Moss and towards Bank End. West of Waitham Hill, parts of Herd House Moss remained unenclosed, while to the north, Waitham Common also lay open, probably largely grazing land; further to the north-west, Bank End Moss also lay open. This is the deepest part of the Moss, and today part is within the Nature Reserve, as is part of Herd House Moss. Further north still, in Broughton, the whole area is divided into strips of peat diggings, which we shall come back to. The other new feature, of course, is the Furness Railway line (extended from Kirkby to Broughton in 1848), which tended to cut off the farm land to the west, isolating Herd House (and leading ultimately to its abandonment).

Twentieth century Angerton

By the early nineteenth century, the Cavendishes had sold out their holding in Angerton - and virtually the whole of the extra-parochial area had come to be held by the Towers family of Duddon Hall; who then in 1902 put their estates up for sale, each farm to be sold separately, as freehold (LA DDHH1/57). The relevant lots for the Angerton part of the estate were Lot 35, Waitham Hill: Lot 36, Moss House, including some land north of Waitham Hill in Bank End Moss: Lot 37, Marshfield, stretching up onto Herd House Moss east of the railway and finally Lot 38, Herd House itself, with mossland to the west of the railway.

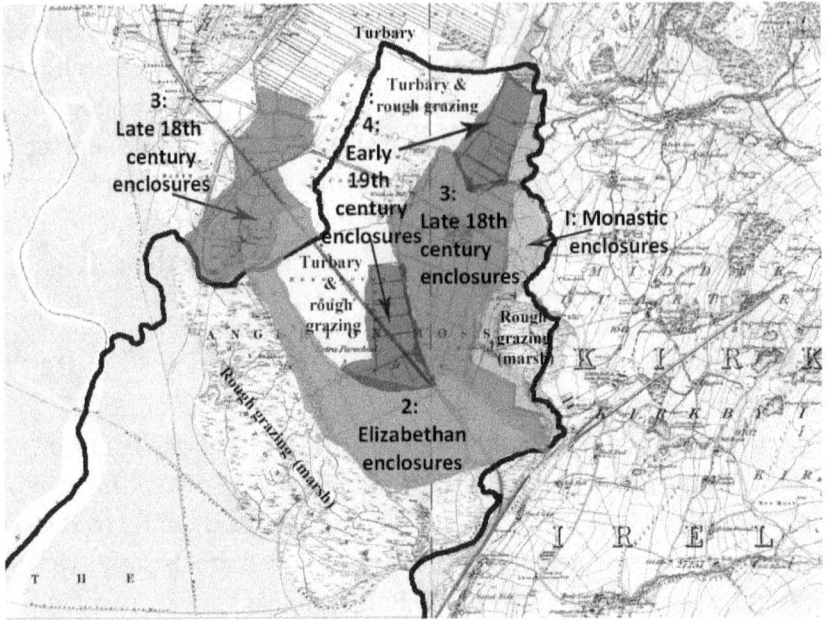

Figure 3. Four phases of enclosure of Angerton Moss (superimposed upon joined extracts of sheets 10 and 11 of the First Edition Six Inch Ordnance Survey, surveyed 1846 and published 1850. (OS map reproduced with the permission of the National Library of Scotland)

Nearly half of the Waitham Hill holding was moss, while the Moss House holding spelled out that it had turbary rights on 74 acres of Bank End Moss. There is no mention of turbary at Marshfield, but it had herbage rights on the Moss; while Herd House had unspecified turbary rights, presumably on Herd House Moss. Herd House Farm, cut off by the railway, had probably been abandoned by this date –but Moss Farm, part of the same holding, has since renamed itself rather grandly as Angerton Hall, and has its very own level crossing, which only goes to the farm.

Oral records tell us that the locals from all these farms continued to cut peat for their fires certainly up to the Second World War – but there was also some commercial peat production going on around the turn of the twentieth century, immediately to the north of Angerton Moss (CCHT, 2015). Although they were outside the historic 'extra-parochial area' of Angerton, the story of these peat diggings is included here as part of Angerton's story. In 1847, the commons of Broughton, to the north of Angerton, were enclosed and divided amongst the owners and tenants (LA AE 4/1). Part of that enclosure involved the former 'moss rooms' of White Moss and Fox Moss, which were then divided up between the various commoners of Broughton into distinctive long narrow strips. Some, such as those strips allotted to the Overseers of the Broughton Poor were around three acres or more, while others were strips of one acre one rood and twenty perches (6,160 square yards). These strips were a furlong long by twenty-eight yards wide, or four local customary perches of seven yards (rather than the statute perch of five-and-a-half yards). This would appear to have been the traditional local standard moss room, here and elsewhere; and the enclosure commissioners had no alternative but to use these strips as their basis, even though they were now using statute measure for their survey.

Similar standards can be seen elsewhere. At Moss House Lane, Much Hoole, in south west Lancashire, former moss rooms, now arable fields, show up clearly on modern aerial photos and also on the Twenty-Five Inch Ordnance Survey of 1892. Here each strip measures four by 100 perches, using the 8-yard perch customary in other parts of Lancashire: but again, the key point is the moss rooms were four local perches wide. A similar example can be seen on the first Edition Six Inch Ordnance Survey sheet 12 (1851), which shows the effect of parcelling out turbary on Ellerside Moss on the Holker estate in the Cartmel Peninsula, in an enclosure event between three local townships – Upper Holker, Lower Holker and Allithwaite. The Ordnance Surveyors have only shown the strips where they fall into different townships, marked on the ground by merestones. These strips are thirty chains long by statute measure – three furlongs – but they appear to be about sixty six yards wide, or probably eight of the seven yard local perches, suggesting again a double width moss room, based on a standard of four customary perches width.

Returning to the subject of Fox Moss in Broughton, the 1892 Twenty-Five Inch Ordnance Survey map shows that since the enclosure of 1847, there had been commercial development in the north part of the Moss. This was the building of a *'Tramway'*, together with a related small terrace of cottages for the work-force, Moss Terrace. By reference to the enclosure map, we can see that the railway lines were built upon the strips acquired at enclosure by one John Garrick – whereupon he, or his successors, began extracting peat on an industrial scale. In the 1901 census for West Broughton, two families are shown as living at Moss Terrace, one of which is headed by Richard Maw, aged 46 who was working as manager of a *'Moss Litter Works'*, assisted by his 25-year old unmarried son, Thomas, described as a *'Factory Hand'*. Interestingly, Richard was from Hull, and his wife and son Thomas were born in Burringham, the Isle of Axholme, in Lincolnshire; while their three younger children were all born in and around Goole. It is probably not unreasonable to suggest that Richard had worked in the commercial peat extraction industry in and around the Ouse, Humber and Trent area before coming over to Lancashire to manage this works. Ten years later, the 1911 Census showed that Thomas was by then Head of Household, describing himself as *'Manager Moss Litter Horse Bedding'*. His brother, like him born at Burringham, Lincolnshire was a labourer at the Moss Litter works - as was John Livesey, a boarder, who was from Millom in Cumberland, just across the Duddon estuary from Angerton, There are no references to any other workers there in the Census, so it was not a very big enterprise.

Moss or Bog Litter had become important in the nineteenth century to meet the demand in towns for horse bedding from the large numbers of urban carters and draymen, not to mention the growth of urban omnibuses and trams. The closure date of this Works is unknown, but seems to have been during WWII. Joss Curwen, a 78 year-old resident of Angerton Moss thus described the works in 2013, from memories of his youth (Rowntree, 2013).

'...there's a track out there and it used to be used for peat graving. That is where they got the peat from for the Bog Litter

Works to load on the railway down to Moss Terrace. They took out several feet of peat.'

And he went on to say

'...Galloper Bridge used to have water gates which were closed to raise the level for the flat bottomed boats going up for peat. They were dismantled to improve the water flow when the Bog Litter work ceased...the remnants of the gate mechanism can still be seen...There was an old man called Marconi...who had a timber built house by Moss Field and a beard to his knees. He was the last Bog Litter worker.'

Thus ended the history of exploitation of the peat in the district, which we can trace back some seven hundred years – but which had probably commenced many hundreds if not thousands of years previously.

References

Ekwall, E. (1922) *The Place-Names of Lancashire*. Manchester University Press, Manchester.

Farrer, W., & Brownbill, J. (1914) *The Victoria County History of the County of Lancaster*. Volume VIII, Archibald Constable & Co., London.

Shannon, W.D. (2010) Adversarial map-making in pre-Reformation Lancashire. *Northern History,* **XLVII** (2), 329-34.

Shannon, W.D. (2012) Dispute Maps in Tudor Lancashire. *The Local Historian,* **42** (1), 2-15.

Shannon, W.D. (2015) Moss Rooms and Hell Holes: The Landscape of the Leyland Dispute Maps, 1571-1599. *Landscape History,* **36** (2), 49-68.

Winchester, A.J.L. (2017) *Cumbria: An Historical Gazetteer*. Regional Heritage Centre, Lancaster.

Printed Maps

Ordnance Survey (1850) First edition Six Inch map, sheets 10 & 11.

Greenwood, C (1818) *Map of the County Palatine of Lancaster taken from an actual survey* W. Fowler & C. Greenwood, Wakefield & London.

Hennet, G. (1830) *A Map of the County Palatine of Lancaster Divided into Hundreds and Parishes From an accurate Survey*, Henry Teesdale, London.

Yates, W. (1786) *The County Palatine of Lancaster*. Henry Boswell, London.

Archival Sources

The National Archives:
TNA DL 42/3, *Coucher Book of Furness Abbey*,
CLXXXIII, Agreement between John de Kirkby Ireleth and Ralph Fitzalan of Kirkby, nd.
CLXXXVIII, Inquisition, 1292, following purchase by Thomas Skillar from Richard FitzSimon of Broughton.
CXC, Grant of Angerton by Thos Skillar to Furness Abbey, 1299.

TNA DL 25/398 Award re Angerton Moss, 1424.
TNA SC 12/9/73 Rental of Furness Abbey c.1533.
TNA E322/91 Surrender of Furness Abbey, 1537.
TNA DL3/48/R5 Commission re Angerton, 1545.
TNA DL 1/137/R12 Duchy Pleadings, Angerton, 1586.
TNA MPC 1/34 Map of Angerton Moss, 1586.
TNA DL 4/83/50 Angerton Moss 1632.
TNA IR 30/18/14, Angerton Tithe Map, 1839, but copied from a survey of 1805.
TNA IR 18/3895, Tithe Agreement 15 June 1839.

Lancashire Archives:

LA DDCA/4/2 Rental, Holker Estate, 1726.
LA DDHH1/57 Duddon Hall estate sale catalogue, 1902.
LA AE 4/1, Broughton Enclosure 1847.

Oral History:
Rowntree, C. (2013) Charles Rowntree oral history interview with Joss Curwen, Angerton resident, *pers com.*

Websites:
Soilscapes: Cranfield Soil & Agrifood Institute:
http://www.landis.org.uk/soilscapes
CCHT (2015): *Draft History of Angerton*, Cumbria County History Trust.
https://www.cumbriacountyhistory.org.uk/township/angerton
Victoria County History
https://www.victoriacountyhistory.ac.uk/counties
Regional Heritage Centre, Lancaster University
http://www.lancaster.ac.uk/users/rhc/index.php

Conference field visit – photographs © Solway Connections Guided Heritage Tours

Chapter 5. Notes on Cumwhitton Moss

David Park

A Short Overview

The following is a short account of my studies on Cumwhitton Moss and its recorded usage. I have read the parchment documents concerning the 1801 enclosure and I have the 1907 paper copy (33 inches x 27 inches) of the enclosure map (see Figure 1). From the parchment document and my copy, I have drawn the shares in Cumwhitton Moss and this map is also included (see Figure 2). I also have a copy of the 1801 enclosure award (36 pages, Carlisle Library A297) and this is informative about the earlier peat-cutting rights and the period following enclosure. The account also draws extensively on the PhD thesis by H.J. Charnley (1973) *The Manor of Cumwhitton Cumberland: a study in historical geography.*

Cumwhitton is a village about seven miles south east of Carlisle and Cumwhitton Moss lies about a half mile south-east of Cumwhitton. Cumwhitton Moss is 108 acres of peatland which has formed in a glacial hollow and is a Site of Special Scientific Interest. Sphagnum moss and other plants grow on Cumwhitton Moss, along with trees including silver birch and scots pine. It is clear that the landscape of Cumwhitton Moss has changed much over the centuries.

A draft copy of the 1603 enclosure bill held by Cumbria Archives states that *'Customary tenants are entitled to.... the common of turbary and to the privilege of pulling furze and fern to be spent and consumed in their customary houses'.* There was also a common right to *'collect the lord's wood'.* Other observations include for example, in 1766, it was noted that there are some *'hazardous pits – outcome of excessive digging'.*

Between 1794 and 1797, Cumwhitton Moss was described as follows (William Hutchinson, *History of the County of*

Cumberland 1794-1797, re-published 1974) *'The mosses are full of wood, oak, ash, hazel; nuts are frequently dug up. From one of the mosses is a strong chalybeate water, this is not singular. The wood buried in this moss lies at a considerable depth…. There are many fine springs of water'.*

From the 1801 enclosure records (Cumbria Archives Q/RE/1/40), it can be seen that in 1771 the Lord of the Manor proposed a private bill for inclosure for Cumwhitton manor. The views of villagers were sought and the manor was subsequently enclosed in 1801. Commissioners were chosen to divide and allot the lands *'to the persons who shall appear to have an exclusive right to the (land)'*. As part of this enclosure, Cumwhitton Moss was enclosed and divided into fifty-three shares. This enclosure is recorded and a map was drawn of these shares, including the shape and size of each share. (See Figure 1 below).

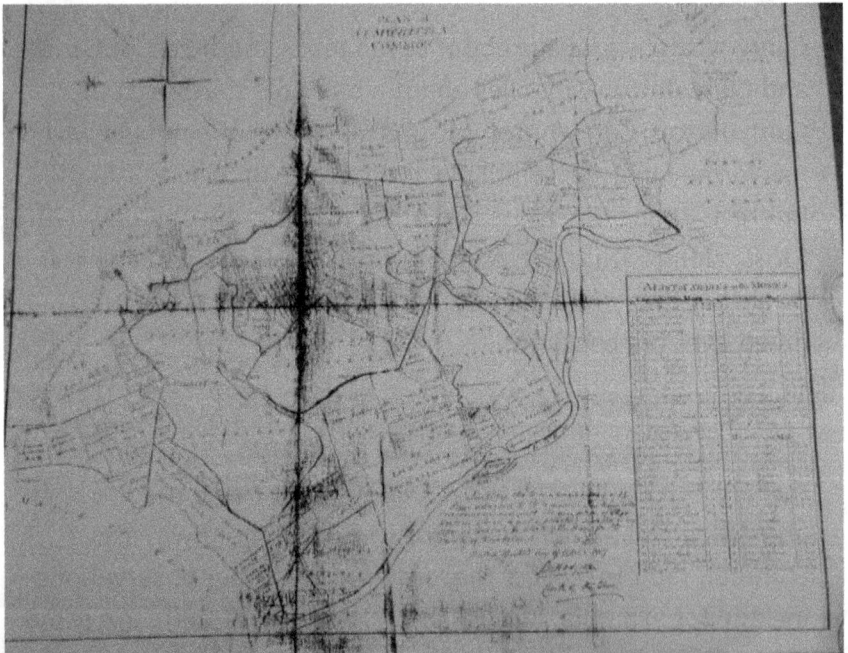

Figure 1. Map to show the allocations of shares in Cumwhitton Moss at enclosure

The map is a 1907 copy verified by the Clerk to the Peace of the County of Cumberland. After the enclosure of Cumwhitton Moss, the enclosure document (Enclosure Acts A297, Carlisle Library) states that it *'shall not be lawful for any person to cut, dig or carry away any turf or sods'*, the previous common right was now only allowed for those people who had been allotted shares. It seems to me that the allocation and sharing of the moss was noted as *'the least selective of processes, in that almost all received a share of less than 2 acres'*. Figure 2 shows a close-up of an enclosure allocation.

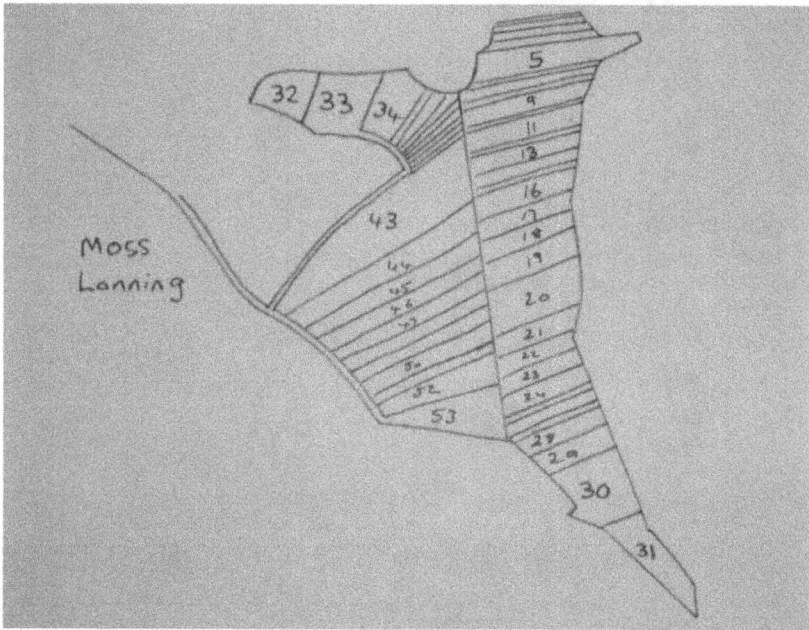

Figure 2 shows a close-up of an enclosure allocation hand-drawn by the author

And, Figure 3, below, is a copy of the enclosure list of allocations.

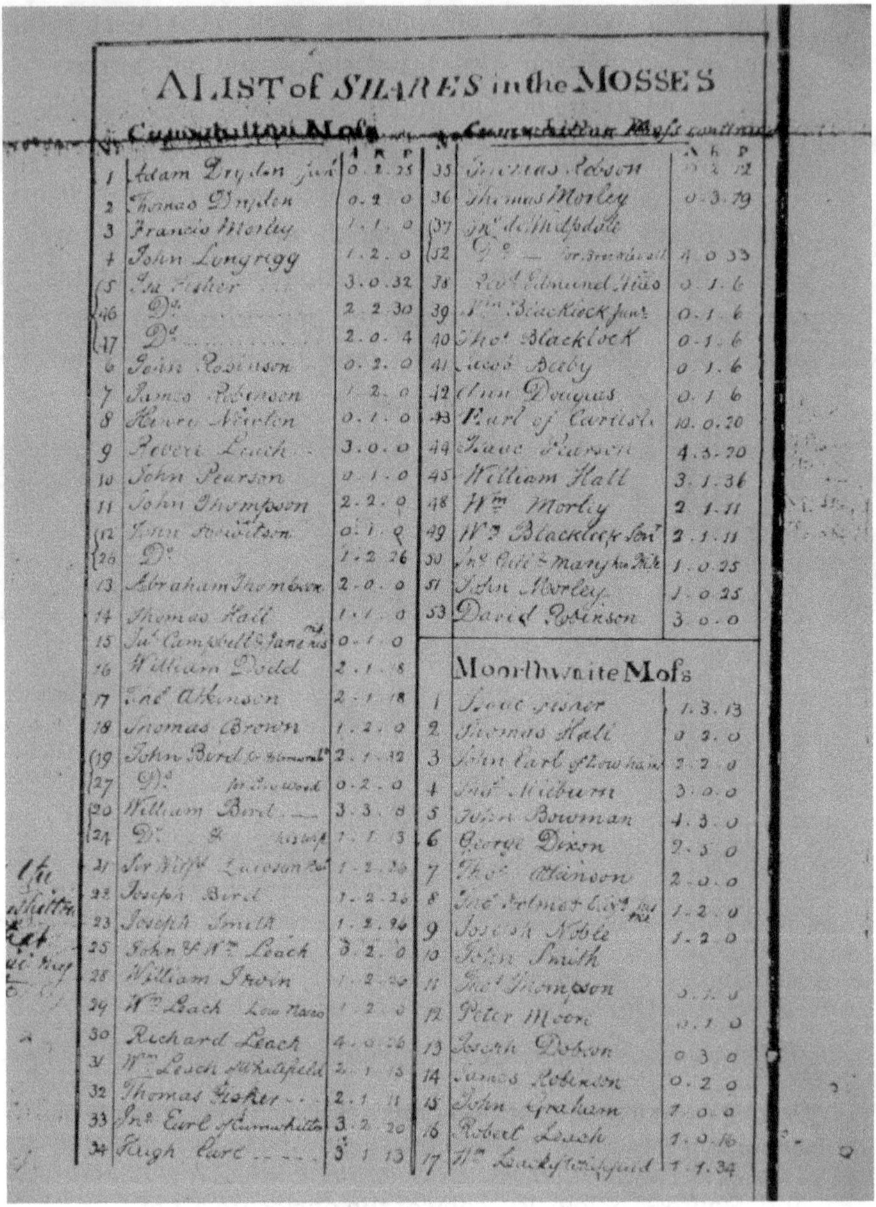

Figure 3. Copy of the enclosure list of allocations

Figure 4. This image of a peat spade is of one that was used on Cumwhitton Moss, it has a new handle fitted with temporary fixing

Figure 5. John Dryden with peat stacks on Cumwhitton Moss at shares 1 and 2

Over time, the digging of peat has changed the landscape of Cumwhitton Moss and areas can be seen where long, deep trenches have been cut and elsewhere there are small, shallow cuts. Nevertheless, the way in which Cumwhitton Moss was worked has changed over the centuries, and by the 1940s only two shares in the Moss were used for peat-digging. The southerly part of the now drier Moss had trees growing on it in the 1940s, and now in 2017, most of the moss is covered by trees. The last peat was dug from the Moss by John Dryden in the 1950s and he is shown in Figures 5 and 6 with peat stacks on Cumwhitton Moss at shares 1 and 2.

Figure 6. John Dryden with peat stacks on Cumwhitton Moss at shares 1 and 2

References

Charnley, H.J. (1973) *The Manor of Cumwhitton Cumberland: a study in historical geography*. Unpublished PhD thesis, Durham University. Available at Durham E-Theses: http://etheses.dur.ac.uk/10022/

Acknowledgments

I am grateful to the owner of the John Dryden photographs for permission to reproduce them. I am also indebted to Carlisle Library and Cumbria Archives for their support in my studies.

Chapter 6. A historical perspective of the Solway Mosses

Francis J. Mawby

Summary

The Solway Mosses peat formation began about 7,000 years ago. Peatlands accumulate a historical record in the form of pollen grains, which in this acidic environment are well preserved. Using pollen grains in peat cores taken from several Cumbrian peatlands Donald Walker wrote his thesis '*The Quaternary History of the Cumberland Lowlands*' (Walker, 1966). Using pollen analysis and historical literature, Walker draws a picture of human influence and occupation of the area. There are few references to the peat mosses before the late 1700s but enough to show that the mosses were being exploited in various ways. From the late 1700s the peat mosses were all subject to 'Enclosure Acts' or Awards, although the Earl of Lonsdale, The Lord of the Manor, retained the peat rights over large areas of Wedholme Flow, Bowness Common, Glasson Moss and Drumburgh Moss.

The mosses were extensively drained following enclosure. The Solway Crossing railway, across the eastern end of Bowness Common, was constructed during the late 1860s.The first record of commercial peat exploitation was in 1858 when a company began stripping Drumburgh Moss and later extended operations to the north side of Glasson Moss. They ceased operations in 1920. Lord Lonsdale leased an area on Wedholme Flow to the Kirkbride Peat Company in 1913 and by 1930, The Midland Peat Company held the lease. The Cumberland Moss Litter Company (CML), formed in 1948, took a lease over Whitrigg Common, and areas of Glasson Moss and Bowness Common. The Midland gave up their lease on Wedholme in 1954 and CML acquired it, closing down their Glasson workings and moving their operation to Wedholme. A company called T. Howlett and Sons seems to have joined forces with CML to process and market the peat. Peat for horticulture became big business in the 1970s and in 1977 Fisons plc acquired Howletts. They sold the business to Levington Horticulture in 1994 who in turn sold to the Scotts Company of America in 1997.

There is documented evidence of cutting peat for domestic use back to the seventeenth century. The 'enclosures' and 'stinting' of peat rights significantly increased domestic cutting in the 1800s but this steadily reduced with the advent of available electricity and coal. There was a resurgence of cutting for domestic use during World War Two but this soon ended after the war years. Only three families were cutting on the Newton Arlosh Awards on Wedholme Flow in 1986 and one person was cutting on the Bowness Parish Award on Bowness Common. The best area of Glasson Moss with a good mire surface and only marginally affected by drainage was designated as a National Nature Reserve (NNR) in 1967. Then in 1981 the Wildlife and Countryside Act made legal provisions for the safeguarding of Sites of Special Scientific Interest and the mosses were all designated SSSI from1982 to 1986. Following designation, the Nature Conservancy Council (NCC), and its successor body English Nature (EN), stepped up efforts to acquire the mosses. The formal declaration of the South Solway Mosses National Nature Reserve (NNR) was in 1994 and Cumbria Wildlife Trust acquired ownership of Drumburgh Moss in 1996. Peatland rehabilitation work on Glasson Moss had begun in 1986 and work has been on-going on all the Solway Mosses since that time. Winter heather burning, which was once a regular occurrence, now rarely happens though the site is vulnerable to wildfires. A fire on Glasson Moss during the very dry summer of

1976 damaged ninety percent of the moss. Archaeological investigations on Glasson Moss in 1996 revealed another use of the peatland with hemp-retting occurring between 400 and 800 years ago.

Keywords: *the origins of the Cumbrian bogs, history of exploitation, Solway mosses, archaeology, hemp, enclosure awards*

The origins of the mosses

During the last glaciation the ice and snow was at such depth that only the peaks of the Lake District Hills protruded above it. However, the whole of northern England district was probably ice-free by about fourteen thousand years ago (Stone *et al.*, 2010). A glacier, moving north-west out of the Eden Valley, deposited material on the Solway Plain creating a gently undulating landscape of low hills or drumlins (examples being Rogersceugh surrounded by Bowness Common and Wedholme Hill, Lawrenceholme and Wedholme House bounded by Wedholme Flow).

The ice and snow returned for a time during a period called the Scottish Re-advance (McMillan, 2004). This brought material from the Scottish hills, particularly Criffel granite, of which some very large boulders have been unearthed. Two of these boulders, both over 1.5 metres high have been utilised as monument stones. The first, located in Finglandrigg Wood National Nature Reserve, carries a fitting memorial to Derek Ratcliffe (1929 – 2005) an outstanding naturalist and former Chief Scientist of the Nature Conservancy Council. He was a key figure in securing the protection of the peat mosses. The second Criffel granite boulder, located at the entrance to Watchtree Nature Reserve (formerly the Great Orton Airfield), carries a memorial to the Foot-and-Mouth Disease episode of 2001.

These boulders alone carry important 'history' lessons of climate, geology, nature conservation and human influence on the Solway.

The landscape created by the departing ice was of low amplitude hills and hollows, and Walker provides evidence to show that the daily ingress of the tide filled the deepest hollows. Changes in sea level and the land rising after losing the weight of the ice through 'Isostatic recoil' cut the lowest areas off from the sea to become freshwater pools that steadily filled with wetland vegetation surrounded by trees. Peat cores taken to measure the peat depth on Wedholme Flow in 1986 frequently encountered tree remains at the interface of the peat with the mineral soil. One such core yielded an intact and identifiable hazel-nut, which collapsed to dust when held gently for inspection. The pollen evidence suggests that over the first few thousand years after the ice retreated, the surrounding landscape became predominately tree-covered. For a time the landscape may have looked something similar to this present-day landscape at the edge of this Canadian glacier (Figure 1)

Figure 1. A modern-day glacier in the Rocky Mountains in Canada

Sphagnum mosses appear in the peat horizon of the Solway Mosses about 7,000 years B.P. forming in the hollows of the postglacial landscape. Sphagnum moss acidifies its environment as part of its growing mechanism and the moss cell structure holds large amounts of water, thus making the perfect niche for itself and other acid-loving plants. Sphagnum does not have roots and grows upward from its apices leaving its older parts in this acidic, oxygen less, waterlogged environment. This environment is an ideal store for carbon and methane gas. Sphagnum grows up and envelopes its own lower leaves and stems and other plant material, which then accumulate as peat. Undrained mires are about 90% water by volume and drainage is very slow with much of the surface water draining through the top 30 centimetres of the living layer. This gives a peatland its special characteristic dome shape leading to the term 'raised mire'. The hollows in the landscape steadily accumulated peat, which slowly enveloped the lower mounds to form a single peat body. The raised mire only receives nutrients from rainfall and the scientific term ombrotrophic or rain-fed is applied. Around the margins of the mire, where it met nutrients and enrichment from the adjacent mineral soil, there would have been a 'Lagg' fen. This fen margin is long gone from the Solway Mosses, drained or removed by agriculture and early peat-cutting. Regular rainfall and a cool climate are required for a mire to accumulate peat. Sphagnum moss can survive short periods of drought but the frequency and duration of rain is the key to its continued growth and peat formation (R. Lindsay pers. comm.).

The western seaboard climate of North West England has it seems long been ideal for peat formation. In an extract from *A History of Cumberland* by Richard S. Fergusson (1890), he interprets the early postglacial landscape thus: "*It was a land of uncleared Forests, with a climate as yet not mitigated by the organised labours of mankind. It is certain that the island, when it fell under the Roman Power, was little better than a cold and watery desert. According to the accounts of the early travellers, the sky was stormy and obscured by continual rain, the air chilly even in summer and the sun during the finest weather had little power to disperse the steaming mists. The trees gathered and condensed the rain; the crops grew rankly, but ripened slowly; the ground and the atmosphere were alike overloaded with moisture. The fallen timber obstructed the streams, the rivers were squandered in the reedy morasses, and only the downs and hilltops rose above the perpetual tracts of wood."*

The morasses, as he describes them, would probably have been prominent in the landscape as treeless wetlands.

Pollen analysis and radiocarbon dating of peat cores from Glasson Moss (Dumayne, & Barber, 1993) suggest that during the Bronze Age the tree canopy cover was almost continuous but tree clearance commenced during the Iron Age. Then *"during Roman Times the landscape was cleared (of trees) round the study sites studied"*; the principal use of timber probably being for constructing Hadrian's Wall. There was also clearance to grow agricultural crops.

By Roman times, Glasson Moss, Bowness Common and Drumburgh Moss had probably coalesced into one large peatland. The Wedholme Flow peat area would have been substantially larger than it is today. West of Carlisle, other significant peat deposits had formed at Salta Moss, Finglandrigg, Orton Moss, Holm Dub and Black Dub. The smaller mosses such as Oulton Moss and Biglands Bog are 'kettle holes' formed where plugs of ice melted out of the glacial clays. Both these examples are at least 15 metres deep. Significant peat mosses also developed along the Scottish Solway and east of Carlisle there are large peat mosses at Bolton Fell, Walton Moss, Solway Moss and Scaleby Moss along with many smaller mosses.

The extent and depth of the mosses, which undrained would have formed low mounds in the landscape, is difficult to imagine nowadays but the deepest peat would certainly have exceeded 17 metres. Even today, the deepest peat on Glasson, Bowness and Wedholme is over ten metres. The survey for the construction of the Solway Crossing Railway over Bowness Common in the 1860s recorded a peat depth of 51 feet (nearly 17 metres) (Edgar & Sinton, 1990). Thorough drainage preceded the construction and was enough to reduce the depth to 10 metres (33 feet), and consolidated the peat to facilitate the laying of the track-bed foundations across the moss. Co-incidentally, a peat depth survey carried out in 1998 showed that the chosen route went over the deepest peat on Bowness Common. Ann Linguard illustrates beautifully the construction of the railway (see *Crossing the Moss* website (www.crossingthemoss.wordpress.com).

Prior to extensive drainage for peat cutting in the 1950s the Wedholme flow 'dome', it was said, blocked the view of the Caldbeck Fells from Newton Arlosh. During this twentieth-century drainage operation Peter Wanning (now deceased and a former Peat Company foreman) told me the mire dropped so fast that each day he had to organise the realignment of the peat railway line. There is no doubt that before drainage the peat mosses would have been remarkable and prominent landscape features. These undrained, peat-forming mosses had a high water content and possibly presented concerns to local people that a bog burst might occur. Indeed, a bog burst occurred on Solway Moss during the night of 16th November 1771, flooding local farms and settlements (McEwen *et al.*, 1989). The cause of the failure was attributed to peat-cuttings that had weakened the natural edge, which gave way during very heavy rain. The fluid peat engulfed many acres of farmland and one or two properties.

References to human exploitation of peat prior to the seventeenth century are difficult to find, however, Hutchinson (1794), tells us that peat and turf was the chief fuel in the parishes of Holm Cultram, Bowness and Kirkbride. He relates that there was 3000 acres of peat moss in Holm Cultram Parish. The intriguing question is when did peat-cutting for fuel begin? Given human ingenuity, it is a reasonable assumption that peat exploitation began when humankind realised dried peat was a good fuel. The peat mosses were extensive, treeless, and perhaps surrounded by wet woodland and during Roman times timber may well have become a scarce resource. One wonders if the monks of Holme Cultram Abbey at Abbeytown, which is close to Wedholme Flow, cut peat for fuel. There is certainly evidence of peat-cutting on the south area of Wedholme Flow (see Figure 10 below.) that pre-dates the early nineteenth-century enclosures. The Holm Cultram Parish records dated 1785 provide documentary evidence of peat stints and the rules and regulations governing their use (Figure 2, County Records Office Carlisle). With the 'enclosures' imminent this could have been a precursor to determining how and to whom stints would be allocated.

The historical significance of pollen

Pollen grains, blown in from the surrounding area and preserved in peat, provide a unique historical record of human occupation and land use in much the same way as silt deposition in a lake (Walker, 1976; Dumayne, 1994). Both studies suggest that wood was a scarce and valuable resource in a largely treeless landscape during medieval times. Investigations by Grainger and Collingwood (1929) reveal quite detailed accounts in the Holm Cultram register and records regarding a Wedholme Wood. In 1675, the sixteen men of 'Wedholme Wood' were charged with procuring and, at times, managing the wood for the maintenance of the sea dyke at Skinburness. The location of this wood is difficult to determine, although the records indicate it was between Sleightholme and the moss, on the hill that determines the northwest corner boundary of the peat. It seems that by the 1700s the wood had all but gone. If that was the only woodland of merit, it demonstrates just how treeless the Solway landscape was in the 1600s. Perhaps pollen analysis might help to locate the woodland site. The only small woodland existing today, the ground flora of which suggests it is very long-established, is on the south boundary of Wedholme Flow on the lower slopes of Lawrenceholme. Wedholme House is another drumlin almost surrounded by Wedholme Flow and that too may have been wooded.

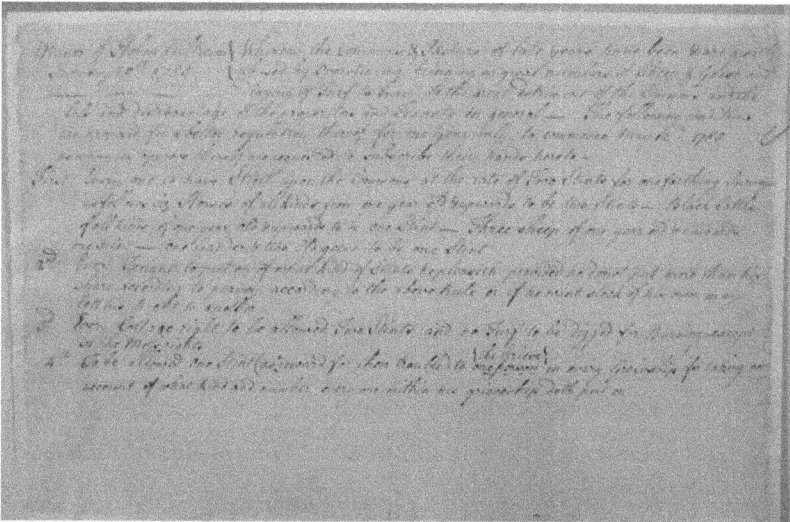

Figure 2. Evidence of established peat stints from Holm Cultram parish records 1785

Hemp-retting

On a dull, overcast day in 1994, I observed features on Glasson Moss highlighted as quite dense circular stands of common cotton grass (*Eriophorum angustifolium*), glistening silvery in the dull light. On closer inspection, the two most distinctive features, almost on the highest point and in the centre of the intact mire, were roughly circular and surrounded by a low ridge dominated by tall heather (*Calluna vulgaris*). Heather is very sensitive to the water level around its roots and only grows well on slightly raised, drier areas. Leading away from the circular areas was a heather dominated linear feature running towards the south-east corner of the moss. The heather and the cotton grass made them stand out from the general intact mire vegetation. Looking round there were other roughly circular areas suggesting that at one time there may have been open water pools on the moss. Talking to colleagues about this brought up suggestions that human excavation or even bombs dropped on the moss during the Second World War might have caused them. The aerial photographs in the English Nature collection provided further evidence of the features. English Nature established the 'Lowland Peatland Project' during the 1990s and instigated peatland training and networking meetings as an integral part of the project. Dr Margaret Cox, an archaeologist at Bournemouth University regularly attended these events to create awareness of the archaeological interest of peatlands. We discussed the Glasson features and a visit to Glasson was organised including a reconnaissance photographic flight. The results were encouraging and in May 1995, Margaret put a small team together to come and investigate the features, which included taking peat cores, extensive walking of the moss and research into historical records at Cumbria County Archives. The results were remarkable (Cox *et al*., 2000) and the analysis of pollen in peat cores revealed that these features had indeed been pools and large amounts of hemp pollen indicated hemp-retting had occurred up to about 790 BP. The use of the natural pools on the moss falls into a period of local production dating from around AD 635 to AD 1630, the principal use of hemp being for making rope. An extensive search of archives

(Chandler, 1996), provided detailed evidence of farmers growing hemp, the fields used to grow it were named and identified on maps, and the seed passed on in wills.

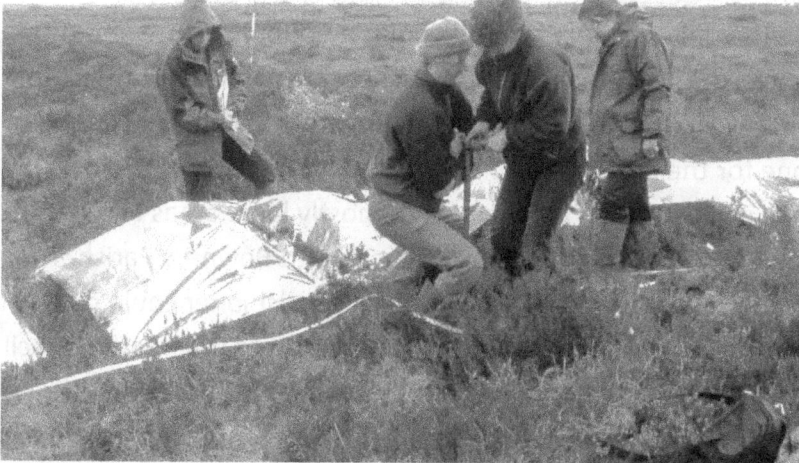

Figure 3. Bournemouth University Team taking the peat cores that gave the evidence for hemp retting

The enclosures of the eighteenth and nineteenth centuries

The Mosses and much of the surrounding land was 'common' until the late eighteenth century. The Earl of Lonsdale held one of the largest 'manors' on the Solway Basin. Within his 'manor' were Wedholme Flow, Bowness Common, Glasson Moss and Drumburgh Moss. The 'enclosure' of common land is extensively documented and debated with many interpretations depending on whether one's sympathy lay with the peasants or the more progressive farmers (Fairlie, 2000). From the early 1800s much of the Solway Basin including the peatlands was 'enclosed'. The mosses were divided into a relatively few private ownerships. On Wedholme Flow three areas of small 'stints' were created for the Parishes of Kirkbride, Newton Arlosh and

Moss Side. The Awards plans, where they exist, show the area and the name of each person to whom a 'stint' was allocated, the Kirkbride Awards (Figure 4, Cumbria County Archives) is a typical example. An Act of Parliament was the usual way of defining the Commons Enclosures. The Kirkbride Enclosure was by an Act dated 1816 and covered the whole parish. It seems to be the only such Act held by Cumbria County Archives for the enclosures on Wedholme Flow. There was an Enclosure Act for the area of Bowness Common in Bowness Parish and possibly one for the area in Anthorn Parish (Cumbria County Archives). The enclosures on Bowness were mostly large areas to a few landowners including an area allocated to Bowness Parish. The 'Parish Moss', was an area where any parishioner could cut peat for fuel, presumably under the supervision of the Parish Council. In Anthorn Parish, there is an area of small 'stints' at the far western end of Bowness Common, the current owners of which live in Cardurnock.

The Enclosure Act for Glasson Moss, assuming there was one, seems lost to records. However, there is a stint map for the southern area of Whitrigg Common, and the 'stint' allocations were to householders in Glasson village. On the stinted area, it seems the peat was shallow and soon cleared down to the mineral soil. The land became rough grassland and acquired the name 'the night pasture'. The 'Commoners' with grazing rights also had grazing rights on the nearby unfenced Whitrigg Marsh, and they herded their cattle to the 'night pasture' every evening. This was to prevent them straying at night and to keep them safe from high tides. A substantial area of deep peat on Whitrigg Common was stinted to several local farmers and this area figured in the early commercial peat exploitation. Within most of the stinted areas on all of the mosses there is one stint of Parish Moss, which seems give those parishioners not allocated a stint somewhere to cut peat for their home hearth.

[This is probably a fuel allowance for the upkeep of the village poor. Ed.] Thus every household at the time had somewhere to cut peat.

Figure 4. Typical stinting Kirkbride Awards (1816 Enclosure Act - Cumbria County Archives)

The 'enclosures' form a very important part of the cultural history of the mosses, with a significant legacy today. Indeed, a whole chapter could be devoted to this topic alone. Further research may reveal how and by whom, stints were allocated and whether stint holders paid the Lord of the Manor for their award. If payment was involved this may explain the difference in the size of each area. Significantly, Lord Lonsdale retained the peat rights to quite large areas of Wedholme Flow and Glasson Moss, and possibly Drumburgh Moss. A stint was probably a valued possession but there seems to be some debate about whether the ownership was to the person or the property. By 1984, some families still held the stints their forebears purchased but many others had changed hands a number of times, to the extent that a small number now have no known owner.

The records for the Finance Act 1910, held by The National Archive Office at Kew, recorded every occupier of land for taxation purposes. These records may show who was cutting peat and where. However, this person was not necessarily the owner because some had given permission to a third party; but this is certainly an archive worth investigating. The notification of Wedholme Flow SSSI in 1986 caused considerable opposition amongst stint owners; the reason being the demand for horticultural peat, which had given added value to the 'stints'. The SSSI notification took away that potential income source. A key issue for many was whether they did actually own the stint or just the right to cut peat? Research by Mr Jock Irving into the 1816 Kirkbride Enclosure Award determined that the award did indeed grant freehold ownership.

Peat fuel was a valuable commodity used extensively for heating homes by the early 1800s. Peat-cutting and drying occurred during the summer with peats ready to be 'lifted' and taken home by early autumn. Old photographs show that peat-cutting was a family affair and probably an important part of village life. The cutting, stacking, and turning of peat may not have been the most popular chore because it would involve running the gauntlet of midges and hordes of other biting insects. However, peat-cutting diminished quite rapidly once coal and then electricity became easily available. Demand probably increased during both World Wars, but after the Second World War, peat-cutting for the hearth quickly ceased for all but a few local people keen to keep the tradition alive.

The registration of Wedholme Flow as a 'Common' under the 1965 Commons Registration legislation was a cause of much consternation to the stint owners. This resulted in Cumbria

County Council holding a 'Commons Enquiry' at which the owners had to provide proof of ownership. Wedholme Flow remains a registered Common, although it seems there are no 'common rights' attached.

The 'Enclosure Awards' are difficult to trace and documentary evidence is not available for a number of areas on all of the mosses, assuming they were all enclosed by Acts of Parliament. The Kirkbride Awards seems to be the only one on Wedholme Flow. There are no documents for the Newton Arlosh Awards, or the Moss Side Awards, or for the areas of larger ownerships on Wedholme Flow. There are documents and maps for the Bowness Parish area of Bowness Common but seemingly not for the Anthorn Enclosure. On Glasson Moss, the only documentation is a map of the awards for Whitrigg Common but nothing for the larger areas of the Moss.

Drumburgh Moss may provide a clue about how other enclosures were carried out as the following extract from the Cumbria Wildlife Trust management plan describes: '*In the 1840's the tithe map for Drumburgh Township shows all the moss area as being owned by Drumburgh Township and was recorded as uncultivated Common or waste land. By 1890 the Ordnance Survey Map shows 45 Turbary Allotments (also known as peat pots or stints). Within their stint an allotment holder had the right to cut peat, and graze cattle, sheep or horses on it. These Commons and the Commoners Rights still exist today. The drainage channels, bulk heads and cut peat faces associated with peat cutting are all physical evidence of historical industry*'.

The enclosures period heralded the first significant drainage of the mosses. The Bowness Common Act defines the drainage in detail, which the new owners were obliged to do to mark their ownership boundary. Bulmers' Directory 1901 relates, '*A large tract of low flat land, forming an extensive moss of several* hundred acres, was drained a few years ago and converted into good farmland. It is known as Bowness Flow, and is situated near the Wampool, extending into the townships of Bowness and Anthorn. The work of reclamation was commenced in 1845 and many miles of drains had to be cut to a depth of 14 to 16 feet, to reach the bottom of the morass'.

Mannix and Whellan (1847), document this in more detail under their Bowness Parish entry: 'A large tract of mossland containing several hundred acres called 'Bowness Flow' situated near the River Wampool in Bowness and Anthorn Townships is now undergoing the fertilizing process of drainage under the superintendence of Mr George Stewart. This great work commenced in 1845 and already many acres of this hitherto unproductive waste are fit for cultivation. The depth of the morass is from 14 to 16 feet and the two main drains, which lead to the Wampool and the Solway are about that depth and from two to three miles in length, which, with the side or cross drains now cut make to about 40 miles. From a farmhouse called Rogersceugh, situated on a small hill in the centre of the moss, is a fine panoramic view'.

Bulmers' Directory says, 'Rogersceugh was farmed by John Harrison Little. Much of the lower part of the farm are deep, drained peat where a former tenant by name John Sibson, in 1878 tried the experiment of training a yoke of oxen to plough and cart on the soft moss land. It proved eminently successful.'

Commercial peat-cutting

Bulmer's Directory further documents the first commercial peat exploitation on Drumburgh Moss.

'About a mile from the village are the Drumburgh Chemical Works, erected in 1856, and extended in 1881 by the addition of a tar distillery. Sulphuric acids, sulphate of ammonia and tar *products are manufactured here. From ten to twenty men are constantly employed'.*

The company was the Guaranteed Manure Company, proprietors R.W. Goold & Co., 32 English Street, Carlisle. It seems they extended their operations to the north side of Glasson Moss along Aikshaw Lonning but to date no records referring to this operation have been located other than the 1895 OS map which shows a 'Chemical Works' (disused).

The first edition of the Ordnance Survey 1895 shows the chemical factory on the small Whiteholme drumlin. The only building belonging to the company remaining today is a house, which is still occupied. The location of the works may have had some connection with the Carlisle to Silloth railway opened in 1856 and closed in 1964. One of the other Whiteholme houses may have been railway property. The company seems to have ceased operations shortly after the First World War and the 1926 edition of the OS map shows Whiteholme Chemical Works (disused). Their workings on north side of Drumbrugh Moss were extensive and removed all of the peat down to the mineral soil.

It is said, by one source dating back to 1610, that Drumburgh village acquired its name thus: (Macpherson, 1892) - of Drumbugh [= Drumburgh] the same writer (Denton M.S 1610) observes '.....it is called Drumbugh of that fenny mire or bog, then full of shrubs and haunt of Bitterns, which people call myre drombles or mire drummles so that Drumbogh signifies the Bitterns Fen'. However, a recent definition of Drumburgh suggests 'the township on the ridge'.

The only remnant of the Guaranteed Manure Company's occupation of Glasson Moss was a derelict chimney, which the Nature Conservancy demolished for safety reasons during 1967.

A woman I met many years ago related a story that the menfolk of Glasson often retired to this area on a Sunday morning for 'a quiet smoke and a crack'.

Commercial peat-cutting commenced on Wedholme Flow when, in 1913, The Kirkbride Moss Litter Company acquired a licence from Lord Lonsdale of a large area of the moss to the south and adjacent to the Newton Arlosh Awards (Figure 5 & Figure 6). The cuttings were set out after the 'Dutch pattern' and it seems probable that Dutch peat-cutting experts set out the cutting areas and drains. The aim was to consolidate the peat sufficiently to cut a peat sod. To achieve this it was necessary to lower the water content from 90% by volume to around 70% by volume. A freshly cut, wet peat sod was about the same cross-section as a house-brick and about one and a half times the length. It probably weighed about the same as a brick approximately 6lb (2.77kg) when wet.

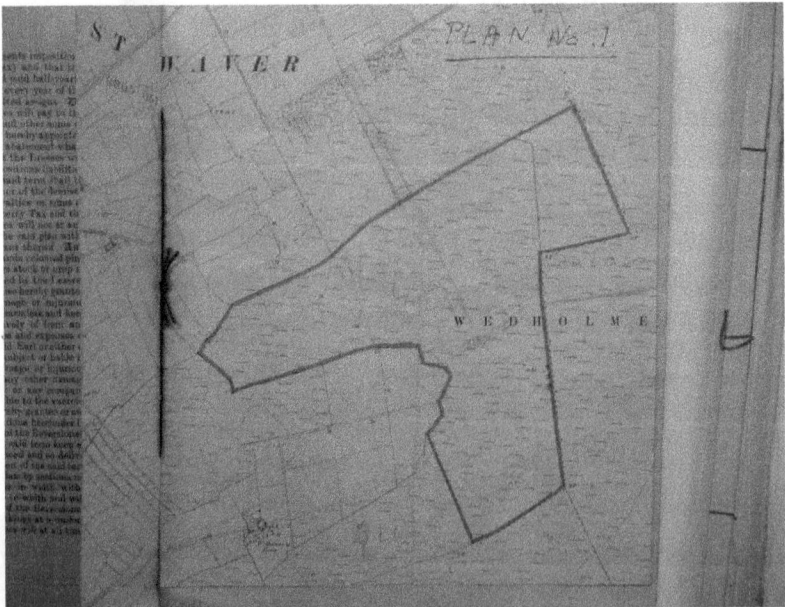

Figure 5: The first 'licence' of parts of Wedholme Flow – Lord Lonsdale to Kirkbride Peat Company

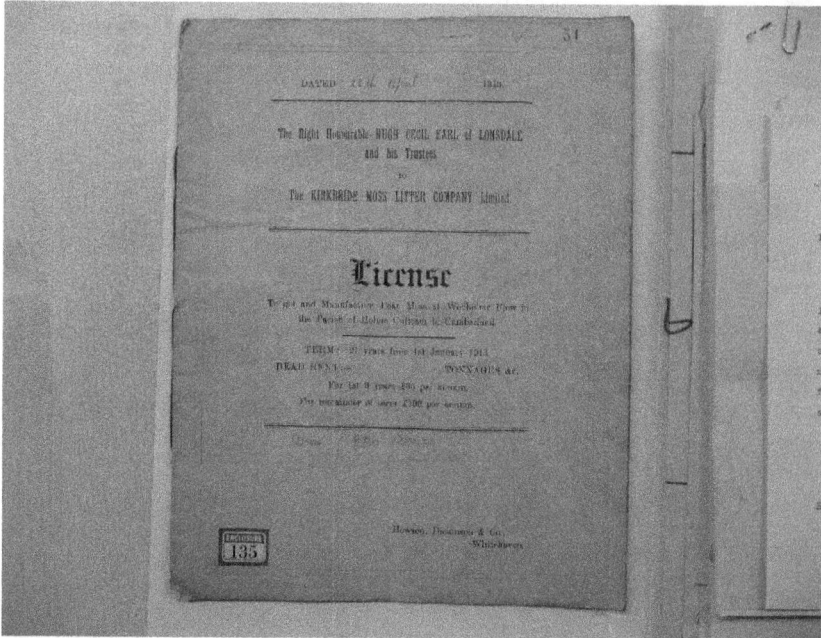

Figure 6: The first licence for commercial peat-cutting on Wedholme Flow

An assessment of the moss topography determined the most effective route for the main drains. From measurements taken on the old cuttings on Glasson Moss, the main drains were dug approximately 44 yards (two 'chains' a chain being 22 yards) apart. Every 88 yards, a nine-yards wide 'stacking' flat was left, where the dried peats were stacked ready to be transported off the moss on a mineral railway line. The next stage was to dig drains perpendicular to the main drains and at eleven-yard intervals. It is difficult to determine the depth of the drains and cuttings from the present-day situation, but the sod cutting machines on Wedholme in the 1980s were taking about half a metre depth. It is reasonable to assume that hand-cuttings were a similar depth. The main drains may have been up to a metre deep. The final step was to skim off the surface vegetation to enable the sod to be cut. The sods were stacked on the edge of the 'cutting' to dry and they were turned to aid the drying process. The vegetation skimmed off the surface was put back into the cutting. It took seven years to remove a full layer of peat between each eleven-yard drain. The dried sods were moved to the stacking flat and transported off-site on the mineral railway. There are no records or photographs of the early means of transporting the peats off site prior to the use of small locomotives; maybe it was horse-drawn carts on the rails.

On Wedholme Flow, on the first area of cuttings from 1913, the dried peat was extracted to the north through one of the Newton Arlosh Awards Stints to the Carlisle to Silloth railway line. This ran between the Moss and Newton Arlosh. A letter to the peat company, dated 1922, notifies them of the closure of the Solway rail crossing. The main drainage of these first cuttings was north alongside the mineral railway, east through the Kirkbride Awards and west through the Moss Side awards.

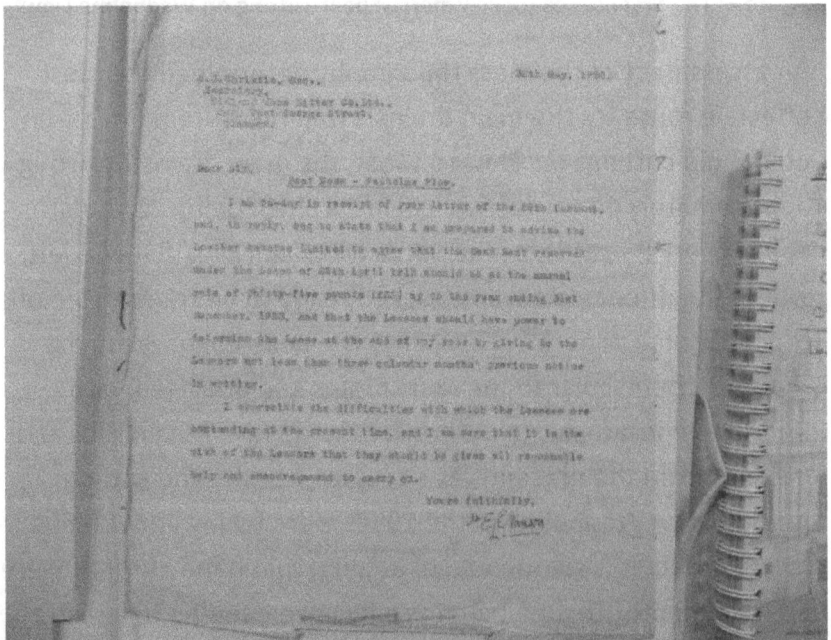

Figure 7: letter from Lord Lonsdale to The Midland Moss Litter Company dated 30th May 1930

A letter from Lord Lonsdale (Figure 7) dated 30th May 1930 refers to a lease to the Midland Peat Company for peat cutting on Wedholme Flow; the Midland working Wedholme until 1954. A Dutchman, Mr Henry Engelen, set up the Cumberland Moss Litter Company in 1947 and correspondence indicates that CML was trying to secure the Wedholme lease. Negotiations with Lord Lonsdale seem somewhat protracted. However, they did secure a lease on areas of Glasson Moss and Bowness Common from Lord Lonsdale and Whitrigg Common Stint owners on Whitrigg Common. They also purchased a substantial area of Glasson from another owner or owners (ref: Natural England, acquisitions record in the Glasson Moss Management Plan). Their peat works was in a field adjacent to the moss on the southwest side owned at the time by Mr Story at Greenspot who was to become their main haulier of the peat products (Mr R. Wills pers. comm. and current owner of this field) (Figures 8 & 9). Cumberland Moss Litter eventually secured the Wedholme Flow lease from the Midland in 1954. They closed the Glasson operation in May 1952 and moved their whole operation to Wedholme (Figure 10.) for better quality peat and to put all of their operations on one site (Mr Pete Wanning pers. comm). Their first peat works on Wedholme Flow was located at the Old Mill site along the Lawrenceholme Lonning. Cumberland Moss Litter had also acquired land by purchase or lease from other owners to hold some 800 acres of Wedholme.

Cumberland Moss Litter drained and established cuttings on some 800 acres of Wedholme and moved their factory to a former Kirkbride Airfield hanger at Low Eskrigg. This hanger caught fire in 1974 when a machine operator, fuelling his machine, lit a cigarette and the air laden with peat dust caught fire (Margaret Sharples pers. comm., who was working there at the time). The whole building, a hemispherical, domed, steel-reinforced, concrete structure built in 1938/9, collapsed inwards. The company who occupied this building was T. Howlett. Margaret relates that Mr Howlett owned a garage in Blackpool but also had a peat moss in Lancashire and, according to Fisons plc history, another in Lanarkshire in Scotland. It is difficult to trace records of Howlett's, or what their relationship was to Cumberland Moss Litter Ltd. They were, it seems, marketing and possibly processing the peat for Cumberland Moss Litter Ltd and the latter was still an active company on the Companies House register into the 1990s.

Reference

ABERLAND NEWS JANUARY 1951 GLASS MOSS PEAT WORKINGS — Glasson Moss peat

LOCAL PEAT IS EARNING DOLLAR

Activities of firm at Whitrigg

A SMALL but lively concern near Kirkbride is playing part in earning American dollars for Britain. It is Cumberland Moss Litter Industry, Ltd., which is exporting peat fibre across Atlantic.

THE PEAT TRAIN ARRIVES

The Cumberland Moss Litter Industry, Ltd., is only one of probably many small industries in and around Carlisle whose efforts in the export drive are little known to the general public. It was more by accident than design that a "News" reporter and photographer this week came upon this active project.

The light railway from the stack yard to the mill has feeder lines to the workings on the common.

The original consignment of two staff men was to find sources of fuel peat for which there has, in recent weeks, been very big demand by Carlisle housewives eager to supplement meagre coal rations.

In the course of their journeys in search of these sources of fuel, reporter and photographer arrived at Whitrigg, near Kirkbride, where the Cumberland Moss Litter Industry, Ltd., is situated.

PEAT FIBRE DEMAND

During an interview with the works manager, Mr Henry Engelen, who is also a director of the company, it transpired that for the past years the firm had been exporting peat fibre to America.

One recent consignment sent to New York comprised 1000 bales of fibre, another for 700 bags despatched to San Francisco, and yet another consignment was sent to Baltimore.

The Kirkbride firm is in strong competition from countries as Ireland, Holland, Poland. Nevertheless, the Americans have expressed a liking for the Cumberland peat fibre.

OF DUTCH STOCK

Mr Engelen comes of Dutch stock. His parents came to this country from Holland in 1904. His father started working among peat when he was twelve years of age, and it was only last year, at the age of 73, that he retired from active work. His son, the present works manager, was born in Lancashire and worked in the moss lands there as well as in East Lothian before coming to Kirkbride some five years ago.

Like his father, Mr Engelen has lived practically all his life within range of the tang of peat. "Until I was 17," he declared, "we burned nothing else but peat in our home, and I still use it on the coal fire."

RUSH FOR FUEL PEAT

The exceptionally cold Winter and the coal shortage are put down as the reasons for the big demand for fuel peat in recent weeks. Housewives have found that they can make their meagre coal ration last a little longer by ...

there is quite a big demand.

PACKING SNAG

At the moment a snag encountered by the firm is question of packing. The Americans prefer the fibre packed in smaller packages from 100 lbs. in weight. Up to the present the firm have been transporting the fibre in bales weighing 3 cwt. and more.

Ways and means, however, are now being devised to meet requirements of the American ...

The fibre is being put to uses in America. It is also being used as fertiliser in gardens and greenhouses, it is also being used as cattle bedding and being mixed with molasses to form a cattle ...

Unloading the baler. A standard pack is seen upright and a lighter pack for export is seen wired up on the trolley.

Figure 8: Cumberland News report January 1951 of peat cutting on Glasson Moss

Figure 9: The 1951 aerial photograph of active peat workings on Glasson Moss, Whitrigg Common, and Bowness Common. The peat works can be seen as two buildings in a narrow strip between the moss and the Port Carlisle road

Figure 10: Aerial photograph showing the southern half of Wedholme Flow in 1949. The extent of the commercial cuttings is shown as are the old pre-enclosure cuttings

By the 1970s, peat for horticulture and amateur gardening was becoming big business. Peat, being a stable, predictable acid medium, proved ideal for making peat compost through the addition of lime and other essential minerals. This attracted the attention of Fisons, the large agricultural fertilizer company and according to reports found on the internet they bought Howlett's out in 1977. Fisons had also acquired a substantial peat resource on Thorne and Hatfield Moors in South Yorkshire and smaller areas on the Somerset Levels. They acquired the Levington brand and introduced 'Grow-bags', designed for tomato-growing, to the domestic gardening world.

Significant financial problems beset Fison's in the early 1990s and through a management buyout in 1994, they subsequently sold the horticultural operation to Levington Horticulture (*The Independent*, July 6th 1994). The horticultural peat business was now under intense scrutiny from the conservation bodies aware that the industry was rapidly destroying the lowland raised mires contrary to both the SSSI legislation designations and the newly issued European Habitats Directive. Much of the peat harvesting had immunity from this legislation because being classed as a mineral it had previously established planning permission.

A small success for conservation came on Wedholme in 1996 when a planning consent (for 20 years from 1976) on the south-east corner expired. The situation with the new legislation meant that the Planning Authority had to take full account of the situation and peat harvesting had to end on this area. Their planning consent on the main area of the peat workings dated back many years and as such had no end date.

An insight into the peat-cutting in more recent times can be found in '*Ask the fellows who cut the peat*' on www.solwayshorewalker.wordpress.com by Ann Linguard.

Nature Conservation and the Solway Peatlands

The 1949 National Parks and Access to the Countryside Act established powers to declare National Nature Reserves (NNRs), to notify Sites of Special Scientific Interest (SSSIs) and for local authorities to establish Local Nature Reserves (LNRs). It conferred on the Nature Conservancy and local authorities powers for the establishment and maintenance of nature reserves; it was the first legislation to recognise the need for National Nature Reserves and SSSIs. The forerunner to the Act was the Conservation of Nature in England and Wales

(Command 7122) 1947 – Report of the Wildlife Conservation Committee. The report gave details of the most important wildlife sites in England and Wales at that time. Under the 1949 Act SSSIs were notified to the local authorities concerned but it seems notification of the actual site owners was fraught with difficulties and legal protection was inadequate to prevent agricultural improvement or development. However, the subsequent Wildlife and Countryside Act 1981 addressed this situation.

The first reference to Glasson Moss being a NNR was in the autumn of 1956. A letter from Mr G.N.C. Swift to Mr E.M. Nicholson (The Director General of The Nature Conservancy) discussed the possible establishment of a NNR on the South Solway. Mr Nicholson's reply (2nd January 1957) was affirmative and mentioned Wedholme Flow as a possible candidate. In particular it was considered that a raised mire site was the most relevant to replace Racks Moss in Dumfriesshire, which had been destroyed by a fire. A letter dated 3rd January 1957 from Dr D.A. Ratcliffe to the Regional Officer Mr R. Elliott stated that he and Ernest Blezard (Curator of Natural History Tullie House Museum Carlisle) considered that Glasson Moss was the best representative example site of the three possibilities in the area (Bowness Common and Wedholme Flow being the other two). This was contrary to Dr F. Rose's view that Wedholme was better. D. Walker was also in favour of Glasson. Ratcliffe and Rose surveyed the sites and on 17th January agreed on Glasson. From thereon, negotiations began in order to secure the best area of Glasson Moss. What may have influenced their decision was the substantial area of active peat-cutting on Wedholme Flow, whereas operations had ceased on Glasson. Bowness Common, one of the largest mosses in England least modified by drainage and peat-cutting, was not considered a candidate because of damage caused by frequent winter burning. Derek

Ratcliffe later described Bowness Common as one of the best remaining geomorphological examples of lowland raised mire remaining in the UK (*A Nature Conservation Review*, 1977). Since this time, Bowness Common has been recovering well following the cessation of regular winter fires and the damming of old drains since 1990, and now has extensive areas of peat-forming sphagnum.

The acquisition of the intact area of Glasson Moss was concluded in 1960 following negotiation with two owners, Cumberland Moss Litter Industries and Lord Lonsdale. The formal declaration of Glasson Moss as a National Nature Reserve was in 1967. Cumberland Moss Litter donated their Bowness Common holding to the Nature Conservancy Council (NCC) in 1984.

A Nature Conservation Review (NCR) (1977) by the Nature Conservancy Council Chief Scientist Derek Ratcliffe was, and still is, the most comprehensive published list of all the important nature conservation sites in the UK. The NCR identified all the South Solway lowland raised mires as important sites along with several other peatlands in North Cumbria.

The Wildlife and Countryside Act 1981 led to probably the most important step towards the conservation of the Solway Mosses with their designation as SSSIs from 1983 to 1986 and the fact that the designation now carried genuine protection. Significantly, the notifications included areas of old and current peat workings with the long-term outlook to achieve the restoration of peat-forming mire vegetation and the protection of the adjacent, intact mire areas to prevent further degradation. The *Guidelines for the notification of Peatland SSSIs* (Lindsay, 1989) considered it desirable to include former mire area, now under agricultural grassland. Had this been implemented the SSSIs would have been considerably larger, especially around Bowness Common. English Nature and the successor body Natural England working in partnership with the RSPB have made some progress to this end. The RSPB has acquired farmland, swapped fields and carried out work to restore agricultural fields over peat back to wetland.

The Peat Campaign

By the mid-1980s, there was a growing awareness that the exploitation of peat for horticulture was unsustainable and a significant threat to the integrity of any adjacent areas of intact mire on peatlands being harvested. In England, much of this 'intact' mire (peatlands with original surface, peat-forming vegetation) is on the Solway Mosses. Furthermore, scientists realised that cutover mires released vast amounts of methane, a 'greenhouse gas' contributing to global warming and climate change. This strengthened the case for the rehabilitation of damaged peatlands. Realising the damage that horticultural peat harvesting was causing to a scarce and dwindling 'natural habitat', ten voluntary bodies set up the 'Peat Campaign' (Barkham, 1993) in 1990. With growing support and public awareness, the Peat Campaign put pressure on the UK Government to stop further peat harvesting on the SSSI sites and urged the horticultural industry to develop peat alternatives and gardeners to use peat free composts. Today, peat harvesting has ceased in England, although peat harvesting continues on some mosses in Scotland. The peat companies have developed some peat alternatives. They say there is not a sustainable and adequate substitute for sphagnum peat. The main source of their peat is now from the extensive peat-cutting which still goes on in the Baltic States, Finland, and Ireland; although the latter has reduced in recent years.

The main driving force following the Peat Campaign came in the form of the European Union 'Habitats Directive'. During the late

1990s English Nature with full Government backing and funding were able to negotiate with Fison's, then Levington's, and finally the Scotts Company to achieve deals. This meant that in 2001 peat harvesting finally stopped on Wedholme Flow, on Thorne and Hatfield Moors in South Yorkshire, and across the Somerset Levels.

The first major success had come in 1994 when Fison's concluded its agreement (first announced at the beginning of 1993) to give 8,000 acres of peatland to English Nature. This included their holdings on Glasson Moss, and Wedholme Flow, Thorne and Hatfield Moors in South Yorkshire, and holdings on the Somerset Levels. The Scotts Company, a large North American Horticultural Company acquired Levington Horticulture in 1997, but continued pressure from the Peat Campaign and English Nature eventually brought the end to peat harvesting in 2002. The deal, worth £17 million, included a three-year rehabilitation plan managed by English Nature from 2004 to 2007, during which Scotts would carry out the rehabilitation work on Wedholme Flow and on Thorne and Hatfield Moors.

Rehabilitation of the Solway Peatlands

Peatland rehabilitation work requires a good understanding of peat hydrology. Raised mires are interesting phenomena in the way peat and the living vegetation skin holds water to create a domed surface. Water movement through the peat is very slow and much of the rainfall drains down the slope through the 30 cm living moss skin. When a drain is cut through peat, the water drains out, the suspended solid material compresses, and the weight squeezes more water out changing the whole topography and hydrology of the bog. Furthermore, air-spaces develop allowing oxygen in and causing decomposition of the peat. The air-spaces that form allow free water movement and drained peat becomes a free-draining medium. Indeed, drainage significantly changes the whole hydrology and topography of a peatland. This is shown clearly by drains dug following the 'enclosures', which had a significant impact at the time and continue to affect the moss topography and drainage to the present day. Rehabilitation of 'cutover' mire requires the re-establishment of a water-table that fluctuates not more than 30cm below the mire surface in the driest periods of weather and has some surface-water pools during the wetter periods. This is a significant challenge in drained peat where the water table may be over a metre below the surface. The aim of rehabilitation is to slow the water movement and restore the water table to within the nought to 30cm range, avoiding large areas of deep standing water, especially on the bare peat of cutover mires. If the water level does not achieve the optimum for sphagnum to establish, purple moor-grass, soft rush, birch, and pine readily colonise. Damming drains in mires with a good vegetation cover or old sod cuttings (where some surface vegetation had been retained in the cuttings), has generally been successful except on areas where steeper slopes have formed. The most difficult areas to restore are those converted to peat milling in the late 1980s.

Work on Glasson Moss commenced in 1986 damming the old 'enclosure' drains. These drains, dug over 150 years ago, were still effectively draining the moss. The objective was to do the work with minimal visual impact. Using a machine on the Moss was not feasible and so volunteer labour was the only option. The dams used polyethylene sheet as an impermeable membrane set in a trench a metre deep and backfilled with peat and a peat dam. Another innovative solution in narrow drains was the use of steel-building cladding-sheet driven into the peat a metre deep. The old commercial peat-cuttings were the next job, and this was a far more complex task. A small tractor with a rear-mounted digger, light enough to travel on the drier, more compacted peat of the old cuttings aided the volunteer workforce together with an all-terrain vehicle for transporting materials. A big step forward in 1990, was to bring in a large excavator, working on timber bog-mats, to dam the old cuttings on Whitrigg Common. This job was not without incident because the machine bogged down whilst travelling onto the site. Retrieved with considerable effort, the machine went on to work very effectively and completed the task. The machine operator recalled that it was 'like working on waves'; the machine movements on the bog-mats rocking and creating waves across the Moss.

On Wedholme Flow, the Nature Conservancy Council acquired the mineral rights of 251 hectares of Wedholme from the Lonsdale Settled Estate in 1987. This included the commercial cuttings on the north side of Wedholme Flow. Fison's lease over the peat-cuttings on this area, which included the original area leased to the Kirkbride Moss Litter Company in 1913 and later the areas leased to CML, expired in 1990. When English Nature gained control of this area rehabilitation work commenced in 1991. Ken Hope Ltd was contracted to carry out the damming work during the winters of 1991 and 1992. His principal machine driver, having previously worked on the Moss for Fison's and Levington's, was the ideal operator on the unpredictable peat terrain. The old sod-cutting method meant that remnants of former surface vegetation remained in the cuttings and the drainage system and layout made damming a relatively easy process. Thus, within ten years, Sphagnum and other typical vegetation colonised a substantial area of these rehabilitated cuttings, although the steeper slopes remained drier. Each year through the 1990s, funded through the Lowland Peatland Project, Natural England acquired more areas of Moss facilitating the damming of many drains on Bowness Common, the intact areas of Wedholme Flow, Glasson Moss, and Drumbrugh Moss.

The rehabilitation work from 1986 to 1994 was the subject of an unpublished dissertation for the Countryside Management Course (F.J. Mawby (1994)) at Birkbeck College, University of London.

The 'Peat Deal' concluded with Scotts in 2002 included a sum of money whereby Scotts, using their staff expertise and machinery, would carry out the rehabilitation work determined by English Nature over a three-year period. This area of Wedholme Flow was converted from sod-cutting to peat-milling in the 1990s; milling being a harvesting method that removes the entire vegetation leaving a bare peat surface with drains at 20-metre intervals leading into very deep main drains. The aim of the rehabilitation scheme was to manage water flowing off a large catchment and re-establish the water table in the peat conducive to the eventual colonisation of peat-forming vegetation.

Rehabilitation work on all of the Solway Mosses continues through to the present day including further enhancement of areas where peat-forming vegetation has been slow to establish. Re-establishment of the lagg fen around the mire margins remains an objective to be realised because it has serious implications for farming and the adjacent landowners.

Acknowledgements: Permission to use the records from the Lowther Estate Trust, Figures 5, 6 & 7. Thanks to Anne Abbs and my wife Shelagh for proof-reading.

Bibliography

Barkham, J.P. (1993) For peat's sake: conservation or exploitation? *Biodiversity and Conservation*, **2**, 556-566

T. Bulmer & Co. (1901) *History and Directory of Cumberland.* T. Bulmer & Co.,

Cox, M., Chandler, J., Cox, C., Jones, J., & Tinsley, H. (2001) The Archaeological Significance of Patterns of Anomalous Vegetation on a Raised Mire in the Solway Firth Estuary and the Processes Involved in their formation. *Journal of Archaeological Science*, **28**, 1-18.

Cox, M. *et al.* (2000) Early-medieval hemp retting at Glasson Moss, Cumbria in the context of the use of *Cannabis sativa* during the historic period. *Transactions of the Cumberland and Westmorland Antiquarian and Archaeological Society*, Series 2, **100**, 131-150.

Dumayne, L., & Barber, K.E. (1994) The impact of the Romans on the environment of northern England: pollen data from three sites close to Hadrian's Wall. *The Holocene*, **4** (2), 165-173.

Edgar, S., & Sinton, J.M. (1990) *The Solway Junction Railway.* Locomotion Papers No. 176, The Oakwood Press, ISBN 0-85361-395-8.

Fairlie, S. (2000) A Short History of Enclosures in Britain. *The Land*, Issue 7, Summer 2000.

Hutchinson, W. (1794) The History of the County of Cumberland volume 2.

Lindsay, R. (1989) *Guidelines for the notification of Peatland SSSIs.* Nature Conservancy Council, Peterborough.

Mannix, & Whelan (1847) *History Gazetteer and Directory of Cumberland 1847.* W.B. Johnson, Beverley.

McEwen, L.J., & Withers, C.W.J. (1989) Historical records and geomorphological events: the 1771 'eruption' of Solway Moss. *Scottish Geographical Magazine*, **105** (3), 149–157.

MacMillan, A. (2004) The Quaternary history of the Solway – revisiting the Scottish Readvance. *Earthwise*, **20**, 23-23. British Geological Survey, NERC.

Macpherson, H.A. (1892) *A Vertebrate Fauna of Lakeland*. David Douglas, Edinburgh.

Ratcliffe, D.A. (ed.) (1977) *A Nature Conservation Review*, 2 Volumes, Cambridge University Press, Cambridge.

Stone, P., Millward, D., Young, B., Merritt, J.W., Clarke, S.M., McCormack, M., & Lawrence, D.J.D. (2010) *British Regional Geology: Northern England*. Fifth edition, British Geological Survey, Keyworth, Nottingham. [Main Late Devensian Glaciation, Quaternary, Northern England].

Walker, D. (1966) The late Quaternary history of the Cumberland Lowland. *Philosophical Transactions of The Royal Society of London*, **770**, (251).

Archival sources:

Cumbria County Archives and The Lowther Estate Trust: DLONS/W/7/4/5/138 documents relating to peat-cutting leases Figures 4, 5 & 6.

Kirkbride Enclosure Awards – Cumbria County Archives File ref QRE/4/56 Figure three.

Register and Records of Holm Cultram (ed.) Francis Grainger and W.G. Collingwood (Kendal, 1929). British History Online. http://www.british-history.ac.uk/n-westmorland-records/vol7

Chapter 7. The former extent of peat-cutting in the Cumbrian fells

Simon Thomas

Cumbria Wildlife Trust

Summary

In the light of the common assumption that upland peatlands are eroding due to natural processes, previous work on the scale of historic impacts from rural land use are highlighted. A consistent pattern has emerged of upland commons being a vital resource managed by and for local communities, with access to peatlands for fuel and other materials regulated just as closely as grazing rights are still allocated to this day. Associated changes to the hydrology have led to habitat change and peat erosion by surface water, though the effects of peat removal cannot easily be separated from the contribution of drainage, vegetation burning and grazing, all of which intensified since the Victorian era. Descriptions of field evidence for peat cutting visible in the Lake District and Yorkshire Dales are presented alongside the documentary evidence.

Keywords: *Peat cutting, erosion, turbary, blanket bog, upland, moorland, commons, Cumbria*

Introduction

Unlike in Scotland and Ireland, where remote cottages still dig peat for fuel, this practice has all but disappeared from the collective folk-memory of the English since largely dying out in the Victorian era. The extent of peat cutting across the English uplands has certainly been documented before though (see Rotherham, 2009). Once you know something of the range of products harvested from remote and long-forgotten sites across the Pennines and Lake District, the landscape starts to reveal its forgotten stories.

Peat-cutting or natural erosion?

In the UK, drainage, removal, farming and burning of delicate peatland systems have been so widespread that I have never seen what I can say with confidence is a pristine example of a blanket bog. Erosion largely dates from the last few centuries (earlier in Ireland), so is firmly within the period of human activity (Evans & Warburton, 2007), making it difficult to separate land management from other causal factors. Several authors report a lack of evidence to indicate that erosion gullies occur on peat where the surface vegetation remains intact, except where gullies migrate and expand from adjacent ground (Tallis *et al.,* 1997; Evans & Warburton, 2005).

The scars in Figure 1 are typical, and include some rectangular shapes cut by hand amongst on-going erosion of the vulnerable bare ground by natural processes. Historical peat-cutting is now known to have significantly reduced the size of upland peat bogs but cannot at present be accurately quantified. It is one of the combinations of factors, largely man-made, that can destabilise habitats and initiate peat erosion. These factors are then exacerbated by on-going 'natural' processes that dissect the blanket of peat (Huckerby *et al.,* 2011). Visits to Ireland or Scottish crofts where peat is still used for fuel will reveal fresh peat banks on the hillsides, alongside older cuttings in various stages of healing over or eroding, and additional drainage channels installed to help facilitate cutting (foreground of Figure 2). Morphologically, these are strikingly like patterns in eroding peat 'hags' on British mountains and moorlands (Figure 1).

Figure 1. Peat erosion scars on Baugh Fell Common, SE Cumbria 2017

Figure 2. Recent hand-cutting of peat, Skye

We can assume that straight lines and right angles are man-made, but this is less clear cut where peat cutting is more haphazard, on hills with variable topography. The onset of water and wind erosion along vulnerable exposed edges permanently removes all evidence of the original peat banks.

Dried peat, the original fossil fuel, is known to have been burned since medieval times, and quite possibly earlier. With tree clearance in the Cumbrian uplands on-going since prehistoric times, the last few remaining ancient woodlands were protected for charcoal production by the sixteenth century. By necessity, dried peat bricks became the main fuel in Cumbria until the increased availability of coal in the Victorian era (Marshall, & Davies-Shiel, 1977). That made peat an economic resource for several hundred years and it would have been sold in the towns. Wooden sledges are known to have been used to move dried peat bricks down from the hills and one can be seen in a Turner etching of a Highland scene (Figure 3), and see Figure 8.

Figure 3. J.M.W. Turner. Peat Bog, Scotland, plate 45 from Liber Studiorum, 1812

It was usual for the tenants of a manor to have the right to dig peat for fuel on the unenclosed fell land, then known as the 'manorial wastes'. This right, known as 'common of turbary' (along with other common rights and a whole host of local issues) was actively policed by manor courts. These courts regulated to try and conserve dwindling stocks (Winchester, 1984). Waste is a misleading term for land that was in so much demand that communities had to agree between them on where each household would gather materials, and limits had to be set. Turbary involved collection of both turves and peat bricks for fuel and as a building material. 'Common of estovers' allowed the collection of bracken (for animal bedding, thatch and the potash industry), gorse (fuel), wood, stone, clay, heather (thatch and fuel), bog moss (building material) and rushes (for lights) amongst other things (Rotherham, Egan, & Ardron, 2008).

The 'Eskdale Twenty-four Book' of 1587 (CRO, D/Ben/3/761) gives a fascinating insight. Twenty four sworn men drew up a detailed award 'for the usage of the common within this lordship with the tenants' goods therein habiting', which gave a framework then used by the manor court and its successors for managing Eskdale commons right into the twentieth century, along with several later additions as necessary. Manor court juries regulated the location of sheep 'heafs', peat 'pots' and bracken beds for each property as well as animal numbers, timing of grazing, peat digging and bracken cutting and restriction of use to houses within the manor, amongst other things. From 1842, the Eskdale court officers included an official *'peat-moss looker'* (Winchester, 2008).

Cumbrian field evidence of turbaries

In a systematic survey for evidence of the cultural and social history of peatland use across an entire English upland landscape, Ardron (1999) concluded that the current extent of blanket peat in the Peak District National Park is the isolated remnants of a once much larger moorland landscape spreading across surrounding hills and out into the lowlands. Peat was systematically removed since at least the medieval period, both by the collecting of domestic fuel and by 'paring and burning' to improve the land for farming. As well as land converted to improved fields, the large swathes of shallow peat with semi-natural habitats also seem to be converted from blanket mire, at least in part (Rotherham, Egan & Ardron, 2008). In a brief tour around the main geographical areas explored so far whilst surveying potential Cumbrian peatland conservation projects, it appears this was also going on high up in the Lake District. For curious readers interested in looking for themselves, all the mosses I mention by name are either on open access land or viewable from public rights of way. Documentary evidence from other sources is also described where available.

Upland fringes of the English Lakes

Many hollows containing deep peat deposits still survive amongst the sheep pastures on the rolling hills bordering the Lake District. Valley mires and basin mires can still be found in the 'low fells' of the south Lakes. Mickle Moss, north of Ings, has two adjacent mires rich in plant and insect life. The northernmost mire has a regular pattern of linear cuttings, long abandoned and now almost completely obscured. Only sharp un-natural vegetation boundaries are now discernible as they infilled once the drainage system clogged. In the southern mire, one isolated island of higher peat suggests that the entire original surface has been lowered by peat-cutting, presumably leaving a small peat baulk for stacking and drying peat bricks. An almost identical situation can be found at Scab Moss, which is now surrounded by the conifer plantation on Claife Heights but has remnant walls across, indicating it was once part of a managed farm landscape.

Several small to medium sized lowland 'raised bogs' also occur on valley bottoms at up to 300m scattered around the foothills of the Lake District mountains. Examination of LiDAR remote sensing data for White Moss (Mungrisdale) clearly shows a whole series of straight-sided manmade steps in the side of the Moss and an un-natural surface topography right across the whole peat dome. Having walked across two other such raised mosses, it is clear that the entire peat surface has been removed from both Shoulthwaite Moss (west of St Johns in the Vale) and Kidbeck Moss (Nether Wasdale). These are on private land and can only be viewed from footpaths on the adjacent slope, as is Newthwaite Marsh, Nether Wasdale (take footpath east from a house called The Gap), a small valley mire with a straight-sided depression in the centre connected to a drain. This feature is surely dug by hand.

Northern Lakes

Above Mungrisdale rises the Skiddaw massif. Here, I visited the site of the oldest scheme in the Lake District to conserve peatlands by blocking drainage 'grips', on the lower slopes north of Skiddaw House. There was now little sign of clear field-evidence of peat faces, but it is known that the whole Skiddaw Forest area was subject to industrial-scale hand-digging of peat as fuel for metal-smelting furnaces during the early industrial development of the town of Keswick. In 1571, a team of 53 men was employed to transport cart-loads of peat down from remote bogs in the surrounding mountains (Rollinson, 1967). Whole fell-sides here have probably been stripped bare of deep peat, and Skiddaw is now notable for having very little peat compared to the surrounding mountains. Just east of the track down towards Keswick from Skiddaw House the flatter slopes of Mungrisdale Common above the Glenderaterra Beck have very shallow peat with some low, straight-sided steps in the surface and occasional small drains. This looks like it has been stripped of peat, and the traces of a stone building may betray the former presence of a small hut for drying peat bricks. On the plateau above, a patch of deep peat has large erosion gullies, apparently spreading from some wider voids in the peat which may well be the sites of former peat workings. Similar patterns can be seen in blanket bog on the plateaux of Caldbeck Common and Uldale Common, on the northern edge of the Skiddaw massif. Wythop Moss, west of Bassenthwaite, also supplied Keswick with fuel (ibid). This large wet basin has only very shallow peat today, with deeper deposits restricted to a few fragments around the periphery.

South from there, the lower more gently sloping moorland in and around the parish of Matterdale have many square miles of poorly-drained rough grazing and forestry on shallow peat. Another speaker at the 2017 Cumbria Bogs conference coincidentally mentioned a thirteenth-century record detailing a local man's right to collect cart loads of peat from Matterdale, at a place name I can no longer find on maps. At the time I was surveying peatland erosion on Matterdale Common, and a commoner told me that they had records indicating that the last place peat was cut was Bruts Moss. High on the track up to Great Dodd from Kates Brow, this remote site is far from any houses, and any walkers who remember having tramped through the gullied eroding peat there will know that it wasn't chosen as a convenient place for locals to gather fuel. More likely it was the last deep-peat left. There is a smaller deposit of deep-peat even higher on the common, just below the summit of Calfhow Pike (660m) on the ridge route up to Helvellyn. This has irregular and steep edges with tall vertical erosion faces higher than a person. Richards (2006) tells a great story, that this:

'......is the result of peat cutting, the old workings of a Mr and Mrs Fisher from Bridge House, situated down in St Johns in The Vale; hence Fishers Wife's Rake on Wanthwaite Crags – a perilous pony-hauled sledgate used to convey the dried fuel'.

That is one of the steepest paths in the Lake District so (notwithstanding that Mrs Fisher was clearly an exceedingly hardy person) if the story is true then many of the eroded peatlands on remote summit plateaux could perhaps also be investigated as the remnants of turbaries. The lower fields ('allotments') mentioned previously, some of which are still trying to turn back to bog, must surely have been exploited first and improved for agricultural use, before people moved uphill across Matterdale Common. Many allotments are now grazed by hardy fell ponies, whose ancestors were no doubt employed to transport the sledge and cart loads of peats back to the village. A small island of deep peat, surrounded by vertical cut faces, remains next to the road at Binks Moss, but is disappearing under Sitka spruce now it has dried out. The lowest slopes of the common, just below the historic route of the Old Coach Road, also have three badly eroded areas of deep peat, Barbaryrigg, Sandbeds and Whitesike Mosses, with large deep voids from which erosion gullies are spreading outwards. Old lichen encrusted marker stones are found on the surface of this deep peat, roughly forming two lines running either side of Barbaryrigg and Sandbeds Mosses to demarcate the turbaries of various local inhabitants, and a grassy ramp leads down from the road to a raised peat baulk connecting it to the eroded area (now temporarily fenced to allow recovery). More subtle evidence of possible turbaries is found right across the common. Occasional low steps, now grassed over, are found in most of the peaty areas. These will either be former erosion or man-made peat faces, and in places they do look distinctly man-made, particularly some regular steps high up Mosedale, and parallel rectangular depressions just north east of the gate to Threlkeld Common. Just north of the common are peaty fields at Flaska, a name that suggests Matterdale Common may well incorporate the '*Flascow Common*' site mentioned by Rollinson (1967) as supplying peat for furnaces in Keswick.

South-western Lakes

The south-western fells of the Lake District have had considerable archaeological attention, since they are known to have been inhabited since Neolithic times. As well as recording finds preserved below the peat, surveys by Oxford Archaeology North (2011) also recorded some peat cuttings, particularly on Cockley Moss and Hesk Fell where large rectangular depressions can be found. East of their transect area, Ulpha Fell common has a whole series of adjacent rectangular depressions in the valley mire of Sike Moss (Figure 4) near the Birker Fell Road, which are naturally flooded and infilling. Eroded peat hags at White Moss and east of Long Crag (replanted by a recent project) stand amongst adjacent areas where the peat surface appears to have been lowered. The south eastern boundary of Ulpha Fell Common has thin peat soils, with many steps in and low, roughly rectangular depressions. This is above the medieval field system east of Crosbythwaite, and I suspect it is faint traces of a turbary dating to that period. The dried out shallow peat may have been shrinking since then.

Figure 4. Quaking bog surface in flooded peat cutting. Sike Moss, Ulpha Fell Common 2010

South of Ulpha Fell, severely eroded deep peat near Kinmont Buckbarrow (recently repaired) was found to be surrounded by extensive areas of shallower peat with vertical scars. This hill appears to have had much of the peat removed. Across the Corney Fell road, the slopes up onto White Coombe common have grassy rectangular voids within the peat that clearly look man-made.

Central Lakes

On Birker Fell Common the blanket bog of Tewit Moss and valley mire of Brantrake Moss have low, straight-sided depressions visible. It appears that partial drainage and cuttings on Brantrake, Foxbield, Tonguesdale and Low Birker Pool Mosses have left straight-sided pools within the mires. The remains of peat storage huts have been recorded around packhorse tracks down to Eskdale from that common too, and peat cutting rights have been documented at Tonguesdale Moss and Low Birker Pool (Winchester, 1984).

Winchester lists thirty-five peat storage huts known as peat scales on boggy plateaux on commons high above Eskdale. He described these upland peat huts in Lakeland as being unique in Britain, except for one other area in Wales since British peat stores were usually much nearer to the houses. In 1982, Mr Baines of How Farm remembering the years between the two world wars, described cutting and stacking peat in May, transporting it to the peat scale after five weeks drying, then transporting it down to burn when finally fully dry in the winter. Each residence in Eskdale and Wasdalehead with common rights of turbary had its own peat scale. Most were simple low dry-stone structures with thatch roofs, constructed between the sixteen and eighteenth centuries. They are distant from any paths, hinting that dried peat bricks were carried down in hand baskets or pony panniers. Some later scales were more substantial Lakeland bank barns with ramps to the top door. These were built near well-made sledge tracks for transporting the fuel to the farmhouses (ibid.). One building was maintained for use by livestock, and more recently the National Trust has renovated some others (Figure 5).

Figure 5. Restored peat scales on Eskdale Common

Amongst the rocky outcrops on the plateau below the climb up to Scafell there are patches of peat that mostly look like they have suffered erosion and widespread peat loss. The unusually detailed records of common rights in the parish mean we know that these were peat banks used by residents of both Eskdale and Wasdale (ibid.). The deep blanket peat of the well-named Quagrigg Moss has been largely removed, leaving cliffs of remaining peat around the edge.

Nearby, the River Esk oozes from the remote upland valley mire of Great Moss. The causeway (Monks Dyke) across was recorded in the Cartulary of Furness Abbey as a turf rampart, formerly with a paling of stakes, built to mark the monks' grazing at Brotherilkeld Farm (Symonds, 1965). East of it is a rectangular depression that might just be where the peat was excavated to build this bank. The southern end of the moss is eroded and mostly lost, reduced to a few lips of peat along the base of rocky outcrops. Walkers to Scafell certainly trample it, but the apparent remains of a cart track at the head of the long path up from the farm suggest it may too have been a valuable source of fuel. Far above it, Pike De Bield Moss, below Bowfell summit, has also lost much of its peat, and could be an even more remote turbary. Mr Temple of Black Hall Farm described that his grandfather had cut peat on Hardknott (pers. comm.), where the many fragments of deep peat hidden amongst rocky outcrops are currently or previously eroding. This was high above the larger valley-bottom mire in Moasdale where I had already noted steps in the surface and drains, suggesting the whole surface had been cut over and lowered until it became too swampy.

The Langdale Fells have been widely surveyed because of the important prehistoric stone axe 'factories'. Peat huts survive on Loft Crag and at Scale Gill, and tracks to Tarn Crag, Pike Howe, Troughton Beck and Scale Gill (National Trust, 2002). The track up the eastern side of Scale Gill continues past the peat huts, to an area of peat cutting. Two other cuttings were found at Tarn Crag and Pike How. Further survey of the area to the east of Stickle Tarn at Great Langdale revealed peat tracks, peat scales, and even clapper bridges, but only a few obvious peat-cutting scars. Furthermore, none were reported from the extensive flatter moorland extending to High Raise either (OA North, 2005). It seems probable that this discrepancy is evidence for peat-cutting being far more extensive than can be deduced simply from the remaining cutting scars.

Documentary evidence of the quantities of peat purchased for a copper smelting furnace at Coniston in the sixteenth century is summarised in Rotherham, Egan, & Ardron (2004). Two other furnaces in the southern Lakes were known to burn peat too. Unsurprisingly, the valley mires and flushes on Coniston Moor have clearly been drained and had much of the peat removed, with small fragments of deep peat left around the margins. They would not be big enough to have supplied the furnace and the local population, so peat must also have been transported in from elsewhere.

Eastern Lakes

The eastern fringes of the Lake District have its largest expanses of blanket bog. Blatchford (2013) studied the 1868 copy of the Longsleddale Inclosure Act (24-Kendal/WDB/35/4/15/9) and was able to map three separate blanket bog areas above Longsleddale where up to five different individuals cut peat for domestic fuel, plus a peatcote (hut). The enclosure awards give details of four different farms having adjacent turbaries on Skeggles Water Moss, and another on nearby White Moss. Apart from twentieth-century drainage (now blocked), the bog surface by Skeggles Water is flat and unusually intact. Even apparently pristine bogs may once have had their entire surface layer completely removed. Around the fringes of the site, straight-sided depressions appear to be yet more peat workings that were not recorded and presumably considerably predate the enclosures. By the track up across the moss from Longsleddale are the walls of a relatively well-preserved peatcote. The Inclosure Award records three smaller turbaries high above the top of Longsleddale near the bridleway over to Mosedale. By superimposing this map onto aerial photography using GIS (Figure 6), I can see that the peat has been removed, leaving a whole series of 'naturally' eroding faces and gullies in the edge of deep peat on the slope above and no trace of straight tool-cut edges remaining.

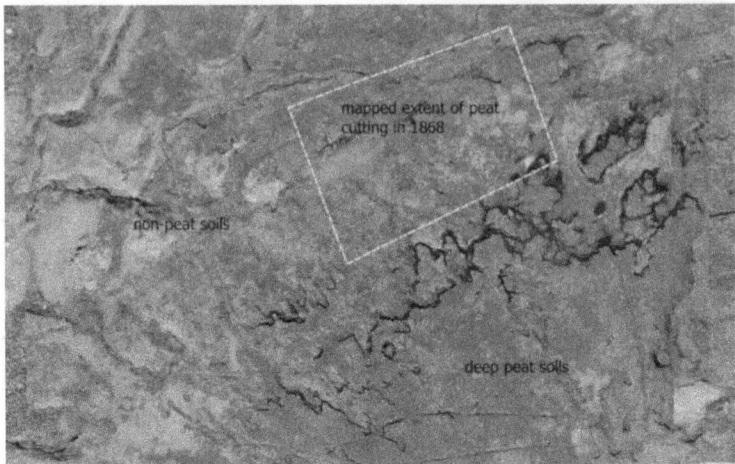

Figure 6. Approximate location of turbaries mapped from 1865 Longsleddale Inclosure Award

Mosedale, and the adjacent valleys of Borrowdale and Crookdale are swathed in blanket bog, with occasional big patches of steps in the peat down into apparent cuttings but confused by active erosion and deep peaty gullies. Farmer Stephen Lord told me that old estate workers referred to the track we were using up from Borrowdale Head as The Peat Road. Maybe the eroding peatland at the top had been exploited by their families. Going north to Mardale Common, there is less erosion but still steep-sided steps in the peat. The village is now lost beneath Haweswater Reservoir, but the National Park Historic Environment Record recorded the remains of a possible turf house of unknown date on the fell above, and mentions an ancient peat-cutting site above. Apparently similar field signs stretch off in every direction, often associated with drains below them. Across the reservoir on Bampton Common the higher ground is also blanket bog. A bridleway follows the Roman road of High Street along the summit ridge, forming a hollow way through deep peat. There are occasional large rectangular depressions with low cliffs at the edge, and more widespread erosion gullies. An aerial photograph of Benty Howe and the slopes above has been published on-line (Historic England, 1935). It shows that the erosion on this section at least was as bad in 1935, before Italian prisoners of war were made to dig drains on the common in the

1940s (Cooke, Garside & Kirkbride, 2013). Clearly, people have been impacting this common long before the well-reported effects of late-twentieth century management on the uplands. Just to the north end is Peatstack Hill, now largely denuded of peat. Here, as elsewhere in Cumbria, substantial tracks can be seen heading up steep hills from farmsteads before apparently disappearing into boggy ground. These can be the most conspicuous remaining clue to the importance these bogs once had. Usually there will be several sunken and parallel zigzagging trackways worn by sledges used to drag dried peat bricks down to the valley.

The blanket bogs of the eastern Lakes extend either side of the A6 road just west of Shap Pink Quarry. Standing by the memorial in the layby, you can see a tall block of peat over the fence, standing high and dry because surrounding peat has been removed. It is now eroding. Across the road under the pylons, are small steps down into old peat-cuttings with swampy conditions where peat appears to be actively regenerating. In 2017, one eroded section had a pool containing the exposed peat-stained remains of an ancient tree trunk. The heavily-drained Wasdale Valley mire between Shap Pink Quarry and the A6 shows no obvious remaining signs of peat removal, but another swampy basin higher up Wasdale Beck has a tall island of dried-out deep peat in one corner, suggesting that several metres of peat have been removed from here in the past too.

Yorkshire Dales National Park

East of the M6 motorway Cumbria has even more uplands smothered in blanket bog peat, in the Cumbrian portions of the Yorkshire Dales National Park, and the North Pennines Area of Outstanding Natural Beauty (AONB). I have spent less time in the latter, so will not attempt to cover that area, save to mention that the manor court records for Alston Moor in 1679 (with a rising population due to lead mining) describe protectionist restrictions imposed on landless cottagers cutting peat on the common without permission from "*adjacent neighbours*". Despite the vast peatlands in this sparsely populated region, there appears to have been conflict over fuel supplies from the fells, at least locally (Winchester, 2000). Peat was widely used as a domestic and industrial fuel in the Yorkshire Dales for many centuries (Rotherham, Egan and Ardron, 2004) and abundant field-evidence can be found there too.

The steep slopes of the Howgill Fells are naturally peat free, but ridges on Tebay and Brant Fell Commons have small patches of deep peat, dried out and dissected by erosion channels. These appear to be the degraded remnants of larger peat mosses. Old zigzagging tracks can be seen incised into steep slopes above many of the farms, mostly now leading to damp grassy ridges with patchy shallow peat. Across Ravenstonedale Common (also patches of peat with drains and peat-cutting scars) above Fell End, Holmes Moss has patchy areas of peat removal, fringed by subsequent erosion. One low step runs in a straight line right across the hilltop. Either side are large upland commons, Mallerstang and Baugh Fell, both with deep peat which has been eroding as long as anyone can remember. The summit ridges have some large areas eroded almost to the bedrock, with sudden steps up to relatively intact peat bog (Figure 7). A hundred years on it is still easy to imagine the busy Edwardian scene in Figure 8 transposed here, with peat carted off to the remote farms of the treeless moorland and limestone country hereabouts.

Some miles south, Kingsdale has an old track along the valley side called Turbary Road, which unsurprisingly leads to Turbary Pasture. Nearby, the green lane from Kingsdale across to Barbondale also gives access to numerous allotments (enclosures) of blanket bog that have both been drained and cut over. Barbondale has a grouse moor, where the bog vegetation is modified by burning to promote young heather growth. This practice is much more widespread as you head east into North Yorkshire and north into the North Pennines. The presence of grouse shooting butts, marked on Ordnance Survey maps, shows that it was formerly practiced at rather more Cumbrian sites in the past.

Figure 7. Black Fell Moss, Mallerstang Common, SE Cumbria (2017)

Figure 8. Peat cutting near Cooks Study, Holmfirth, West Yorks mid-1900s

Conclusions from field evidence

When assembled here together, my selective examples actually give an almost continuous coverage of the main peatland areas of the two National Parks in Cumbria. A consistent pattern has emerged of upland commons being a vital resource carefully managed by and for local communities, with access to peatlands for fuel and other materials once regulated as carefully as grazing rights are still allocated today. It is also clear that there were competing economic pressures to overexploit peat stocks as fuel for homes and furnaces in the growing towns and industries. A survey of sources across the north of England (Huckerby *et al.*, 2011) reveals similar histories in Lancashire.

References

Blatchford, B. (2013) Map of turbary rights from Longsleddale Inclosure Award of 1865, Unpublished.

Evans, M.G., & Warburton, J. (2005) Sediment budget for an eroding peat-moorland catchment in northern England. *Earth Surface Processes and Landforms*, **30**, 557-577.

Evans, M.G., & Warburton, J. (2007) *Geomorphology of Upland Peat* Blackwell Publishing, Oxford.

Mayfield, B. & Pearson, M.C. (1972) Human interference with north Derbyshire blanket peat. *East Midland Geographer*, **12**, 245-51.

National Trust (2002) *An archaeological monitoring report for the Great Langdale Valley*, unpublished report.

OA North (2005) *Stickle Tarn, Great Langdale, Cumbria: Archaeological Survey Report*, unpublished report.

Oldfield, F. (1970) *The ecological history of Blelham Bog National Nature Reserve*. In: Walker, D. & West, R.G. (eds), *Studies in the vegetational history of the British Isles*. Cambridge University Press, Cambridge, 141-57.

Rollinson, W. (1967) *A History of Man In the Lake District*. J.M. Dent & Sons Ltd, London.

Rotherham, I.D., Egan, D., & Ardron, P. (2004) Fuel Economy of the Uplands: the effects of peat and turf utilisation on the uplands. In: Whyte, I.D. & Winchester A.J.L. (eds) *Society,*

Landscape and Environment in Upland Britain. *Society for Landscape Studies Supplementary Series*, **2**, Society for Landscape Studies.

Rotherham, I.D., Egan, D., & Ardron, P. (2008) Lessons from the past – a case study of how upland land-use has influenced the environmental resource. *Aspects of Applied Biology*, **85**, 85-91. *Shaping a vision for the uplands.*

Rotherham, I.D. (2009) *Peat and Peat Cutting*. Shire Publications, Oxford.

Symonds, H.H. (1965) *Walking in the Lake District*. W&R Chambers, Edinburgh.

Winchester, A.J.L. (1984) Peat storage huts in Eskdale. *Transactions of the Cumberland and Westmoreland Antiquarian and Archaeological Society*, new ser., **84**, 103-15.

Winchester, A.J.L. (2000) *The Harvest of the Hills: rural life in northern England and the Scottish Borders, 1400-1700*. Edinburgh University Press, Edinburgh.

On-line sources

Cooke, K., Garside P. & Kirkbride, B. (2013). *Bampton Commons Community History Project 2012-13* for 'Commons Stories', University of Lancaster, Bampton and District Local History Society: *https://www.bampton-history.org.uk/members/Bampton-Commons-Community-History-Project-Report.pdf*

Historic England (1935) *EPW048623 ENGLAND. Benty Howe, Whelter Bottom and environs, Bampton Common, from the north-east, 1935*. Aerial photograph: *www.britainfromabove.org.uk*

Huckerby E., Cook J., Quartermaine J., & Gajos P. (2011) *UPLAND PEATS - MANAGEMENT ASSESSMENT Revised Final Report - Volume 1 Version 3*, Report to English Heritage by

Oxford Archaeology North, Lancaster:
https://oxfordarchaeology.com/articles/171-upland-peats-management-assessment

Lake District National Park Historic Environment Record 6620, Mardale Common Hut:
https://archaeologydataservice.ac.uk/archsearch/record?titleId=2886871

Richards, M. (2006) Great Dodd from Threlkeld:
http://www.bbc.co.uk/cumbria/content/articles/2006/02/20/park_and_stride_great_dodd_from_threlkeld_feature.shtml.

Acknowledgements

Figure 2. View from the lower part of the Quiraing (cropped) Published by Berit: www.commons.wikimedia.org under a Creative Commons Attribution 2.0 Generic licence.

Figure 3. The Art Institute of Chicago. Made available under a Creative Commons CC0 Public Domain dedication.

Figure 6. Aerial Imagery ©2018 Infoterra Ltd. and Bluesky.

Figure 8. Postcard 'Peat Cutting, near Cooks Study, Holmfirth' (collection of Ian Rotherham)

Figures 1, 4, 5, 7 ©Simon Thomas

Chapter 8. The Great Moss: The story of Lochar Moss, Dumfriesshire - Its Origins, Archaeology, History &Wildlife

Peter Norman

An introduction to the Great Moss

The headwaters of the Lochar Water rise on Watchman Moor, a few miles to the north of Dumfries, at an altitude of little more than 200 metres above sea level. They flow a mere 30 kilometres before entering the Solway Firth near the formidable fortress of Caerlaverock Castle. Unlike the nearby River Nith, which makes itself known, sometimes angrily, as it passes through the middle of Dumfries, the Lochar Water sneaks quietly past the back door of the town. As it drops below the 30-metre contour near Locharbriggs, the valley bottom broadens and dark peat begins to stain the water. From here, almost to the Solway, the modern Ordnance Survey map is dotted with numerous place-names associated with peatlands: Mossdale, Tinwald Moss, Manse Moss, Downs Moss, Carnsalloch Moss, Dargavel Moss, Redhills Moss, Collin Moss, Craigs Moss, Racks Moss, Ironhirst Moss, Townfoot Moss, Holmhead Moss, Longbridge Muir, and Cockpool Moss. Others, forgotten since the mid nineteenth century, include Watslacks, Braidmyre and Cranberrie Moss or the Goosedubs. These are the Lochar Mosses. These were once even more extensive, giving the impression of a single vast expanse of wild wetland carpeted in mossy reds, browns and greens and known simply as *The Lochar Moss.*

Of the individual mosses that survive, some are little more than names on the map. But most, including several very large sites, still support some characteristic peatland plants and animals. The Lochar Moss rarely features in the everyday lives of Dumfries' current human residents, but this has not always been the case. From the earliest days of human occupation to the mid-twentieth century, it has been both asset and liability, providing a defensive barrier, food and fuel, but hindering transport and agricultural development. It could not have simply been ignored as it is today. The Reverend Jacob Dickson in his 1791 account for the Parish of Mouswald speculated that *"because this district was nearest **the great moss**, called Locharmoss, perhaps it might have originally been named Mosswold...which from the situation of the church, has been a striking object since time immemorial, and almost the whole of which (for 12 miles in length, and full 3 English miles in breadth in some places) is seen by the observer in one view."*

Origins

> *"First a wudd, and syne a sea;*
> *Now a moss, and aye will be"*

This old local rhyme is still occasionally heard from those who live on the edge of the moss. It has been in use since at least the eighteenth century, a time when knowledge of geological time periods and events was limited to those described in the Bible, notably the great flood. Evidence that Lochar Moss was once a sea was frequently unearthed during peat digging, in the form of sea sand known by the local name of 'sleetch'. But the New (1834) Statistical Account of Tinwald parish goes further, stating that Tinwald Isle, a place eight miles inland, was said to be marked upon an old Dutch chart as *"the safest and most commodious harbour for shipping in Scotland"*. The 'Isle' still appears on some modern maps but we now know that although the Lochar Moss was indeed once a tidal inlet as far inland as the 'Isle', this pre-dates both Dutch navigation and the Old Testament by several millennia.

Today, motorists on the A75 or passengers on the Carlisle to Glasgow railway line cannot fail to notice the River Nith as they cross it on substantial bridges, but Lochar Water remains anonymous. Not surprising, given that the crossing could probably be completed on foot by an athletic run and jump; yet in places the Lochar floodplain is more than four kilometres wide. In the early twentieth century, local geologist Robert

Wallace became convinced that *"the Lochar rivulet"* that *"meanders about in an aimless fashion through a vast wilderness of peat"* could not have created such a wide valley; and he was probably right. During the last ice age more than 12,000 years ago, glaciers flowed southwards towards the Solway. As they retreated, they deposited sands, gavels and clays over the sandstone bedrock and created a low ridge, separating the valley of the Lochar from the much larger Nith valley to the west.

Much of the Lochar floodplain is barely above sea level today, but buried deep within the peat is a record of past environments that shows that 7,000 years ago it was an inlet of the Solway. At that time, sea levels were 5 to 9 metres higher than today meaning much of the current Lochar floodplain would have been sand-banks and saltmarsh, inundated by tides depositing fine, impermeable marine silts and muds twice a day. Even today, fragments of marine shellfish and other marine fauna can be found in these deposits. Above the reach of the tides, the low-lying land would have been no drier, flooded by freshwater as each incoming tide blocked the outflow of the Lochar Water. Gradual deposition of alluvial silts restricted percolation of water into the ground, slowly exacerbating the water-logging.

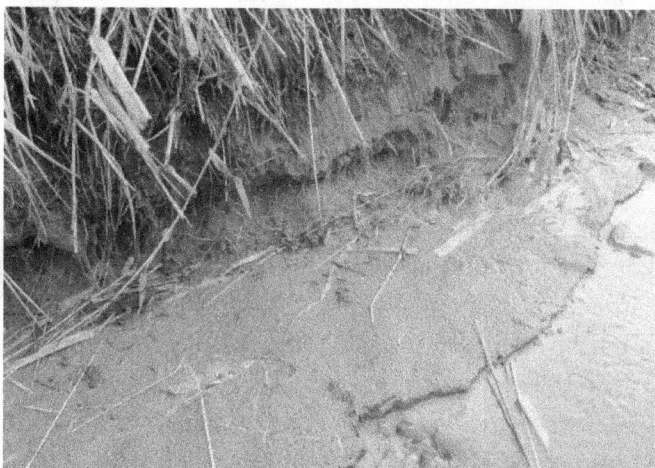

Figure 1. 'Sleetch', the layer of marine sand that underlies the peat; here seen at the bottom of a drainage ditch at Tinwald. © Peter Norman

Seven thousand years ago, sea levels began to fall from their post glacial highs and the land rose in response to the lost weight of ice. Coastal processes in the Solway Firth created an extensive barrier of gravels and saltmarshes across the valley mouth, significantly reducing the speed at which the Lochar emptied the lower regions of its drainage basin. By 6,600 years ago, this barrier became so extensive that the Lochar basin, quite suddenly in geological terms, ceased to be an inlet of the sea. It did however continue to flood regularly, initially with brackish water but, as the barrier became more substantial, freshwater spilled out across the floodplain, creating ideal conditions for peat formation. The barrier continues to grow today, stretching almost completely across the valley mouth and causing the Lochar Water to turn sharply eastwards before entering the estuary.

Peat is essentially part-decomposed plant material. The shallow ponds and lakes that would have formed in the undulations on the valley floor were colonised by dense reeds, sedges and horsetails, with other plants such as water-lilies in deeper water. As these died, they were prevented from fully decomposing. Still or slow-flowing cool water has low oxygen content and any present would have been rapidly used up by plant growth thus restricting decomposition by fungi, bacteria and invertebrates. Semi-decayed material accumulated, infilling the open water from the margins to the centre. Behind it, new and different vegetation dominated by sedges and especially by bog-mosses colonised the water-logged, nutrient-poor conditions.

Bog-moss, or *Sphagnum* to give it its scientific name, is a remarkable plant that can store up to twenty times its own weight in water. It releases organic acids that stop almost all decomposer organisms. It grows slowly but around half of its annual growth remains un-decomposed. Furthermore, each of the thirty or so British species has its own growth form; some forming hummocks up to 50cm high. As the tops of such hummocks become too high to take up water and nutrients from below, they are reliant on dew, mist, fog, rain or snow.

Eventually these hummocks join up so that the whole surface is raised into a great dome composed of 2% semi-decayed *Sphagnum* and 98% water. Cloaked in a protective mantle of living moss, the whole 'raised bog' system depends for its existence on precipitation.

There is good evidence from Priestside Flow, a raised bog under 5km east of the Lochar Mosses that peat formation was well advanced by 8,900 years ago. It was temporarily interrupted by two periods of rising sea levels, but then continued unbroken from about 6,800 years ago until today. It is reasonable to assume that the Lochar Mosses followed a similar timeline, the current peatlands probably being around 6,500 years old. Over this period, the dome grew some 5 to 6 metres above the surrounding land. If the depth of the original lake is added to this, the maximum depth of peat may have been more than 8 metres. Remarkably, it has been estimated that it may take a raindrop up to 90 years to reach the bottom of the peat after falling on the top of such a dome.

The other claim from the old Lochar rhyme, that the moss was 'once a wood' is supported by the fact that the remains of trees including oak, 'fir' (pine), birch, and hazel have frequently been found buried in the peat. Several of these were noted as of great size and in such a sound state that they were used by local carpenters. One such example, a Scot's Pine tree "found in Lochar Moss, standing in an erect position" was described in detail in the *Transactions of the Antiquarian and Natural History Society* (1866):

"*While casting peats this year in a part of Lochar Moss called the 'Syke', in the parish of Torthorwold, and the property of Sir Alexander Grierson, Bart., Mr. John Kerr, farmer, came upon the tree in question, which attracted his attention from its unusual position. In preparation for the party visiting the spot, Mr. Simpson had labourers employed, and the peat removed. The trunk of the tree was uncovered until the root was reached, spreading out upon a grey sandy subsoil. The tree, a Scotch fir, had grown from this soil, and now stood in its original position, the peat having formed around it...The depth of the peat surrounding the stem of the tree was fourteen feet, and exhibited at the bottom a very compact texture. It there contained the remains of jointed reed-like plants, showing that in the early formation the place was marshy. Among this compact peat seeds of plants were abundant, with the remains of various insects. The latter were so preserved as to enable us to distinguish the corslets and wing-cases of carabidous species, and the wing-cases of a species frequenting aquatic plants, Donacia, easily recognised by the beautiful sculpture upon them, here finely preserved."*

Scientific examination of the wood, which was found to be saturated with an oily fluid, was inconclusive with regard to the date of the tree. However, it was suggested that because a Roman coin (see next section) had been found close by, this might give a clue as to the age of the peat and therefore the date that the tree was engulfed. Rather conveniently, this would also tie in with the widely held belief that most Scottish peatlands were created following the felling of woodland by Roman armies. We now know that the remains of the aquatic plants and the wing-cases of the reed-beetles were a much better clue, demonstrating that although some trees, possibly even a wood, undoubtedly once grew here, they were submerged by wetlands, then by peatlands, around 4,000 years before the Roman coins were lost.

However, wood fragments have also been found in a borehole at Sandyknowe and radio-carbon dated to approximately 7,500 years ago. This suggests trees were present here before deposition of marine sands. The old rhyme may have got the sequence of events right after all!

Discoveries: "Lochar Moss and its buried treasures"

'History of the Scottish Nation' is an ambitious subject matter for anyone to research and publish, even more so in the nineteenth century. In 1886, the Reverend J.A. Wylie attempted such a venture. Despite running to three volumes, he never got beyond the early Middle Ages and, unsurprisingly given his occupation, he concentrated on ecclesiastical history. However, he still considered the *"Lochar Moss and its buried treasure"* worthy of a mention on his contents page, alongside St Ninian, Macbeth, and David I. Many treasures and other artefacts had been discovered in the peat prior to 1886 and this continued to be the case into more recent times. The finds cover virtually all historical periods.

At the end of the last ice age some 10,000 years ago, just as the first stages of peat formation began in valley of the Lochar, the first humans were beginning have impacts on the landscape. Few traces of these people have been found, but peat is better at preserving evidence of their existence than other environments. Discoveries on the Lochar Moss have revealed snippets from the lives of people around the peatlands; lives involving skilled craftwork, travel, commerce and conflict. Even more remarkable is that along with preserving clues to their lives, occasionally the peat also preserved the remains of the people themselves.

Bog Bodies

The remains of several human bodies have been found buried in the peat of Lochar Moss. Near Torthorwald, the *"leg of a child cut off (as it appears) by the patella"* was discovered in the seventeenth century. It was covered in a substance described as *"stickish like"*; the tibia and *"fibia"* [fibula] being *"inhosened in a casement like the black bark of a tree"*, which was possibly the remains of musculature. This account may be one of the earliest records of a possible bog body in Britain.

Sometime around 1863, a stone coffin was discovered in the Moss near Tinwald Downs. The report in the *Transactions of the Natural History and Antiquarian Society* does not include details of the find and no further information has been traced. Presumably, as there was no mention of a body, the coffin was empty.

A leather sandal was unearthed from a depth of 2.7m (9 feet) in 1709, and in 1871 a pair of sandals was discovered, this time still being worn by their owner. The *Letter Books of the Natural History and Antiquarian Society* record a skeleton that, in addition to the sandals, was wrapped in or associated with a piece of cloth. These were exhibited at a meeting of the Society in 1872, and an exhibition in the Mechanic's Hall, Dumfries on 7th July 1873 noted the following: "*Item 194. Shoes found in 1871 on skeleton in Lochar Moss...Item 195. Being a piece of cloth in which the body had been buried.*" The eventual fate of these items is not known, but they could be the same human remains as those listed in 1888 to be amongst the contents of Dr T.B. Grierson's private museum in Thornhill. The description was of a "*portion of Woollen Cloth, enclosing human bones found in a moss at Rochs (Racks Moss), in the parish of Torthorwald, Dumfriesshire.*" His collection was dispersed in 1965, local material going to Dumfries Museum, but this cloth and the bones were not listed in the material transferred.

The most recent discovery of human remains came in 1947. The exact location is unknown but it was reported at "*the work in progress on the peat at Lochar Moss*" but "*not on the old track that traverses the Moss*". Most likely it was in Dumfries or Mouswald parish. The discovery was made by a German prisoner of war digging peats at a depth of 1.8m (6 feet) from the surface, though some 0.6m (2 feet) of soil and vegetation had already been removed from the surface by this time. He washed the remains and showed them to the Medical Officer at the Barony Camp, who identified them as a skull and the fourth and seventh vertebrae of a human. There was no sign of peat disturbance above the place where the skull was found, indicating that it had been gradually covered by the peat, rather than buried. The bones were later examined by Dr W.C. Osman Hill of Edinburgh University, and found to belong to a young adult. Deposited in the Anatomical Museum in Edinburgh University they have since gone missing.

As all human remains associated with the Moss are now lost, their exact ages cannot be calculated, but are likely to be between the late prehistoric and medieval periods. John Pickin, former curator of Stranraer Museum, made a study of Scottish bog bodies and compared these to the famous 'Lindow Man' discovered in a peat bog in Cheshire in 1984 and now in the British Museum. He is believed to date from 1,900 to 2,200 years ago.

Neolithic, Bronze and Iron Ages

Made from greenstone volcanic tuff from the scree slopes of the Great Langdale and Scafell areas in Cumbria, a rough-out of a stone axe from Lochar Moss was from the Neolithic or early Bronze Age. It may not seem too odd that this axe-head was found on the north side of the Solway, having been manufactured a short distance from the south side. However, this was unlikely to result from pure accident. The Langdale axe factories produced thousands of stone heads distributed all over Britain with around a third of all English axe-heads from this location. Now in Dumfries Museum, the Lochar axe was probably part of a substantial trade.

Figure 2. Bronze Age flint tools

Bronze Age flint arrowheads or spearheads may have been acquired as valuable items for trade, rather than just as practical hunting weapons. Several have been found, including arrowheads with barbs and stems. These items were purchased for National Museum of Antiquities of Scotland in 1902 and are still in their collections.

The Statistical Account for Tinwald Parish notes that, before 1791, *"a canoe of considerable size and in perfect preservation, was found by a farmer, when digging peats four or five feet below the surface, about four miles above the present flood-mark; but it was destroyed before any antiquarians had heard of it."* This boat was discovered at a depth of between 4' (1.2m) and 5' (1.5m) during peat-digging. In all likelihood, it dated from the Bronze Age.

A second dug-out canoe from the Lochar Moss can be attributed to the Bronze Age with certainty. The stern section was found in September 1973 during mechanical drainage operations on the Old Course of the Lochar Water, an area previously extensively drained. Its exact position in the peat is not known since its significance was only recognised by a visiting geologist after being placed to one side by the excavator driver and left for some time. It was quickly removed to a storage tank in Dumfries to prevent further deterioration, before undergoing conservation work in the National Museums of Scotland laboratories. It was then returned to Dumfries Museum where it remains on display today.

The maximum surviving length of this canoe is 2.42m, and the width varies from 0.75m to 0.81m. It was hollowed from a split (presumably half-sectioned) trunk of oak probably measuring about 1m in diameter. Oak was the favoured wood for such canoes, since it splits along its grain easily. Further shaping was carried out with polished stone axes and flints. The starboard side was badly damaged but the port side survives in part to its full height, suggesting an internal depth of 0.23m.

Canoes such as this were used for wildfowling and fishing and it may have been steered and propelled through the wettest parts of the Lochar Moss using poles. A radio-carbon date puts it in the early Bronze Age, sometime around 2,143 or 2,183BC; the earliest securely dated log boat in Scotland. Given that part of the boat broke away during digging operations, the front section is presumably still somewhere in the Moss.

**Figure 3. Remains of Bronze Age dug-out canoe from Lochar Water ©
Dumfries & Galloway Museums**

Perhaps the most famous, and certainly one of the most
beautiful archaeological finds from the Lochar Moss is an early
Iron Age torc discovered by a labourer cutting turf, about 2
miles north of Comlongan Castle. This beaded collar, made from
cast and sheet bronze, was found dismantled and missing two
beads but in otherwise perfect condition. It was inside a bronze
bowl standing on three hewn stones. A sign of nobility or high
social status, it would have been worn at the neck, either with
the decorated plaque or the beaded section showing at the
front. The bowl is typical of many Iron Age pots used for eating
or drinking. It measures 16.2cm in diameter by 7.5cm in
maximum height and was apparently lathe-turned, the rim
strengthened by being turned in on itself. The deliberate placing
of these objects together suggests some sort of sacred offering.
They were first exhibited in May 1846, shortly after discovery,
and are now held in the British Museum, London. These items
can probably be attributed to the first or early second centuries
AD and the torc is an uniquely sophisticated example of its type.
It has been associated with La Tène culture, a late Iron Age
culture originating from central Europe. A sword with maker's
stamp and a wooden shoulder-yoke was probably part of the
same find; these are now in Dumfries Museum.

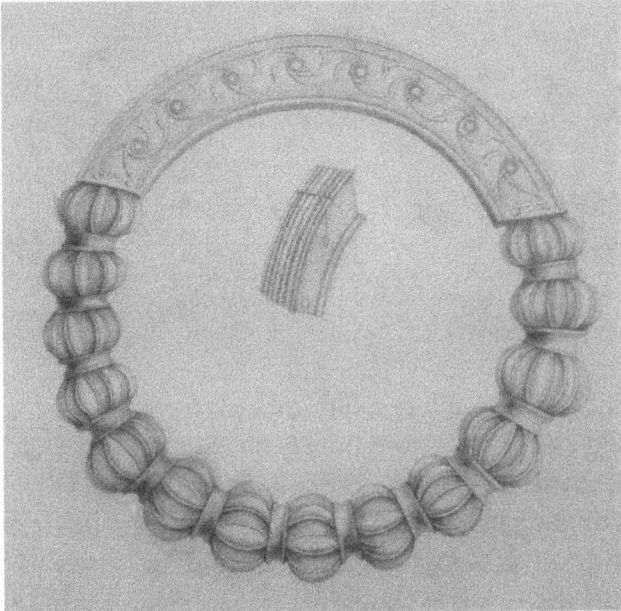

Figure 4. Iron Age torc from Lochar Moss

A jet ring found in Lochar Moss in 1840, is now also in Dumfries Museum and has been dated to the Iron Age. Perhaps the former owners of these items were itinerant traders, or perhaps they lived on the edge of the moss. At Cockpool there are the probable remains of an Iron Age settlement, known as Moss Castle. Only earthworks remain, consisting of a low stony mound with an encircling outer ditch, some 10 to 14m (35 to 45 feet) wide, with an entrance on its eastern side.

Figure 5. Iron Age jet ring © Dumfries & Galloway Museums

The rather grand name arises from a theory that the 'Manor of Cockpule' was built on top of the settlement; the original seat of the Murrays from about 1320 to 1450 AD when they built nearby Comlongon Castle. But, as there are no remains of stone buildings on the site, the theory cannot be confirmed.

Romano-British

A number of discoveries from the Lochar Moss are associated with the Roman period, and it is possible that there may be a link between the bog bodies and high-status Roman metalwork found there. A small onyx Roman seal engraved to represent Mars Gravidus, was found at the 'Syke' and exhibited in 1846; a worn silver family coin of Fufius Calenus, struck 82 BC, was recovered from an old woman in Liverpool around 1866 and traceable to one of 15 or 16 silver coins found about 1.8m (6 feet) below the surface, during peat-casting in the 'Syke'. A Roman bronze cup embellished with dancing figures and a wreath of vine leaves around the neck was found at a depth of about 9 ft around 1852. As with many other finds, their current whereabouts is unknown.

Medieval

During the medieval period, the Lochar Moss appears in the written records. It may have even given its name to a medieval family, with Robertus Loghirmoss, mercator de Scotia, granted safe conduct to travel in England in 1373. Such documents, along with remains of several medieval buildings around the edge of the Moss, allow an historical account of the Moss to be constructed. Archaeological artefacts from the medieval period provide additional clues to life in and around the Moss.

A piece of Red Deer antler found in dredged silt on the bank of the Lochar Water was fashioned into a tool. It may have been lost by a resident of the nearby medieval moated site at Stanhope. Stone balls, of both granite and greenstone, were also dug from the peat in the nineteenth century. These may be of medieval origin and used as missiles, but they may be from an earlier date and status symbols, part of a game, fishing weights, or of unknown use.

In 1768, several gold and silver coins of the *"Jameses of Scotland and some of the Henries of England"* were found at *"Locherness, a mile from Dumfries"*, almost certainly a reference to the Lochar Moss. Given that they were discovered inside a cow's horn, it seems unlikely that they were simply dropped accidentally by a passing baron. However, having not been seen since 1845 the cause of their loss remains a mystery.

Transport and Defence - "the impassableness of the great Lochar Moss"

The Lochar Moss still presents an obstacle to transport, but until recent centuries such a barrier could be turned to advantage; a formidable defensive barricade against potential attackers. The position of the lowest crossing point over the River Nith (now Dumfries), was considerably strengthened by the vast array of marshes and bogs that surrounded it. The Lochar Moss once extended to its very edge, on land upon which much of the modern town is built. Only a low, bushy hill rises above the morass. Perhaps this was the 'Dum Freash' upon which the town was founded and that gave the town its name.

The Gill Loch was the deepest part of the marsh known as the Watslacks or Wetslacks, which along with Cranberrie Moss, drained into the Lochar Water via the Dow Lochar. The view eastwards from the hill upon which St Michael's Church now stands would have encompassed much of this wetland. Today it includes the fire station, schools, shops and many houses. The modern names Lochvale, Gillfoot, Rashgill, Gillbrae Road, Mosspark Crescent, Gilloch Avenue and Moss View are clues to the wetland location. Ancient plant and animal remains discovered in Dumfries attest to the area's wet nature. *The Transactions of the Natural History and Antiquarian Society* (1880) report the following:

"Mr F. W. Grierson brought under the notice of the meeting a remarkable deposit of peat recently exposed near Rae Street when excavating foundations for new houses, together with a number of specimens picked up there a few days ago by Dr Gilchrist and himself, consisting of various pieces of birch and other woods, one of which was penetrated by a green fungus, supposed to be Peziza crucifera. *There were also in the collection*

various nuts, seeds, and debris of coleopterous insects (beetles)."

And...

"A paper was read from Dr Gilchrist on 'The Peat Formation at Newall Terrace, Dumfries,' in which the writer enumerated the objects found in the peat by Mr F. W. Grierson and himself, shown at last meeting, and mentioned that the coleopteral remains consisted of the elytra (wing-sheaths) of the beetle Douacia comari, which still occurs in Lochar and other mosses in the neighbourhood." (This is the Water-lily Reed-beetle, now named *Donacia crassipes*.)

Another large moss which lay north of the Wetslacks was Braidmyre covering the area that now stretches from the Railway Station to the Edinburgh Road and east to the junction of the Moffat and Lockerbie Roads. Later, as the loch shrank, it became known as Black Loch, reputedly a burial ground for victims of the cholera epidemic that claimed more than 420 lives in Dumfries in 1832. Remnants of this loch still survive in the Scottish Wildlife Trust Nature Reserve at Ladypark.

Castles and Towers

The Lochar Moss provided a most formidable barrier to anyone wishing to enter or leave Dumfries by land. There were narrow access routes to the north and south, and these have long been guarded by defensive structures. At the southern tip of the 'Craigs' ridge that separates Lochar Moss from the Nith, lies the Roman fort of Ward Law. Little is known about its use, but it is likely to have been selected as a key strategic location, commanding views of the sea and land for miles around. A short distance away is the most famous of the defensive structures built to guard the access routes around the Moss, Caerlaverock Castle. Caerlaverock consists not of a single castle, but two castles, positioned to protect the narrow route between the coast and the bog. The upstanding stone remains date from the 1270s; the earthworks of the older castle from 1220, but on a site probably occupied from the ninth century. 'A Survey of Scottish Topography, Statistical, Biographical and Historical', edited by Francis H. Groome published between 1882 and 1885, described the position of Caerlaverock Castle:

"...near the mouth of the Nith, 7 miles SSE of Dumfries. Its site is low ground, not many feet above high water mark; was naturally surrounded with lakelets and marshes; and is sometimes called, by the country folk, the 'Island of Caerlaverock.' It naturally possessed considerable military strength, of the same kind as that of many old fastnesses situated on islets or in the midst of great morasses; it has always possessed also the strong military defensiveness of near environment by the surging tides of the Solway and the Nith, and of the impassableness, by an army, of the great Lochar Moss, or of being so situated that it can be approached, even at many miles distance, only along the sort of isthmus between the upper part of Lochar Moss and the Nith; and it, therefore, was in the highest degree, likely to be selected at an early period as a suitable place for a great artificial fort."

Of course, in July 1300 Edward I did manage to gain access to Caerlaverock Castle, but he took no chances, bringing eighty-seven knights, more than 3,000 men and several siege engines to take a castle held by only sixty defenders. In all likelihood he and most of his army crossed the Solway by one of the several 'waths' or fords that are located in the inner Solway; the remainder of the army and several siege engines came from various locations in England and Scotland, including Lochmaben. But it would not have been possible for an army of such a size to cross the Lochar Moss and they must have passed along the coastal route to the south, crossing the Lochar Water near the castle. Perhaps such an army was not just a safeguard against the perceived strength of the castle, but also a precaution against taking such a risky route.

The day after the siege of Caerlaverock ended, Edward had no difficulty in reaching Dumfries, located where the southern and northern Lochar routes converge with the lowest crossing point on the River Nith (originally a ford, but bridged from 1430 to 32). A timber motte, now in the park known as Castledykes, was established here in the twelfth century, possibly by King William the Lion who granted the town royal burgh status in 1186. It was rebuilt in stone in the 1260s but at the time of Edward's arrival after the Siege of Caerlaverock, captured in 1298 it was already in English hands. Presumably the English troops had approached by the less well-defended northern approach around Lochar Moss.

Other strongholds encircle the Lochar Moss. To the north-east is Amisfield Tower. The current building, which still stands to its full height, was remodelled in the nineteenth century but includes a substantial part of the tower built in 1600, and it is possible that a castle has been on this site since the twelfth century. Not as formidable as Caerlaverock, Amisfield may have provided protection for the northern route round the Lochar Moss.

Other fortifications along the 'Wald' ridge or eastern edge of Lochar Moss hint at existence of early routes across the bog, roughly following the line of the small track from Collin that was certainly established by the eighteenth century. Such routes were likely to have been usable only by medieval foot traffic, and only when conditions allowed. At Tinwald, there are earthworks suggesting a motte and bailey castle. Perhaps this guarded the start of a route. Just to the south, at Torthorwald the evidence is a little stronger. The ruined shell of the castle that currently stands on the site dates to the late fourteenth or early fifteenth century, being occupied until 1715, but earthworks show that a twelfth century motte once stood here, and traces of a medieval sunken track have also been found heading in the direction of the Lochar Moss. Despite this, tradition has it that when Robert the Bruce left Torthorwald on his way to meet John Comyn in Dumfries in 1306, a meeting that resulted in Comyn's murder, he went round by *"the skirts of the Tinwald hills, thus making a considerable circuit along the upper extremity of the moss."* However, given that it was

February, even if a route across the moss existed, it is unlikely that it was possible for Bruce to use it.

A little further south of Torthorwald, the remains of the motte and bailey castle at Rockhall offer little evidence for a route across the moss. However, a charter by William de Brus to Adam de Carlyle, issued between 1194 and 1214, suggests such a route, specifying that Adam's tenants of Kinmount should have access to market at Dumfries through Rockhall. Had Adam's tenants skirted round the north of the moss, they would still have almost 10 miles to go, almost half of their journey, after passing Rockhall. Able to cross the Moss, this was reduced to less than 5 miles. Interestingly, the castles at Tinwald, Torthorwald, Rockhall and a sixteenth century tower at Mouswald, all sit directly adjacent or very close to the only modern roads that traverse the Lochar Moss.

Near the southern end of the Moss, as well as Caerlaverlock, is Comlongan Castle. More a defensive tower than castle, it was built in the second half of the fifteenth century by Cuthbert of Cockpool. Like at Collin, there is also some suggestion of routes across the southern part of the Moss, with the Caerlaverock Parish entry in *'A Topographical Dictionary of Scotland'* (1846) stating that the Lochar Water *"flows through an extensive moss, which prevents all communication in that quarter, except in the driest months of summer, and then it is passable only by pedestrians."*

On the 'Craigs' ridge, on the western side of the moss, Isle Tower, also known as Lochar Tower, Bankend or Isle Castle was protected on three sides by the bog. It was recommended for strengthening in an English military report of the 1560s, its strategic position in the whole of Nithsdale considered so important that Caerlaverock was relegated to the position of *"a garrisone assistant."* The tower ruins still stand today and can been seen a little way upstream of Bankend Bridge, but only small fragments of the wall remain intact.

Figure 6. Isle Tower with Longbridge Muir beyond. © Peter Norman

To the north, the base of one of the walls of the tower at Barnkin of Craigs remained until the nineteenth century. The name 'Barnkin', 'Barnekin' or 'Barmkin' was a Scots word for a medieval defensive enclosure around a tower house. No trace remains, though the place-name transferred to the farm.

Roads and Bridges

'A Topographical Dictionary of Scotland' (1846) notes that in 1617:

"James VI, after his accession to the crown of England, visited his ancient dominions, and, passing through Dumfries, remained for one night in the town...The house in which the king lodged was built by a poor labourer who, having found a large treasure while digging peat in the Lochar moss, took a journey to London, where, in a personal interview with the monarch, he was allowed to retain possession of it, and advised to build a house, in which the king promised to lodge when he visited his Scottish dominions."

Whether this story is true or not, James VI of Scotland (James I of England) certainly did visit Dumfries on 4th August 1617, the last day he ever spent on Scottish soil. His return to England facilitated by the construction of a bridge at Bankend to assist his passage around the southern edge of Lochar Moss. This is clearly not the relatively short current bridge that was constructed in 1812; perhaps it was the long bridge that gave its name to Longbridge Muir.

Further improvements to roads through and around the Moss continued over the centuries. In 1724, another bridge over the Lochar was built from the proceeds of the sale of the estate of a tobacco merchant called Pirrie. The Dumfries Archive Centre still holds a copy of Provost Edgar's report, stating that the burgh of Dumfries had gone to considerable expense to build a high road through Lochar Moss, which needed a timber or stone bridge over the Water of Lochar, and calling for a joint committee to discuss this. This was because money was still owed by the burgh to the shire for building this road. General Roy's map of around 1755 shows the road out of Dumfries with a bridge, but little in the way of a road east of the Lochar Water. Perhaps the dispute still had not been settled to enable completion of the route.

By 1757, the 'high road' was included in a survey of the route from Sark Bridge to Portpatrick, for the purposes of *"speedy and certain communication...especially with regard to the passage of troops from one Kingdom to the other"*. The report recommends use of the 'present road' to Dumfries, suggesting that it had been completed by that date. Improvement works were carried out by parties of troops under the direction of military engineers from 1773, though the detail of works on the Lochar section is not known. Sometime afterwards, maintenance of this section transferred to a turnpike trust, with a toll cottage in Collin.

Nevertheless, the passage through the Lochar Moss had a reputation for danger well into the nineteenth century. Sir Walter Scott's novel 'Old Mortality', first published in 1816 included a passage about the Moss:
"My friend, the Rev. Mr. Walker, told me, that being once upon a tour in the south of Scotland, probably about forty years since, he had the bad luck to involve himself in the labyrinth of passages and tracks which cross, in every direction, the extensive waste called Lochar Moss, near Dumfries, out of which it is scarcely possible for a stranger to extricate himself; and there was no small difficulty in procuring a guide, since such people as he saw were engaged in digging their peats - a work

of paramount necessity, which will hardly brook interruption. Mr. Walker could, therefore, only procure unintelligible directions in the southern brogue, which differs widely from that of the Mearns. He was beginning to think himself in a serious dilemma, when he stated his case to a farmer of rather the better class, who was employed, as the others, in digging his winter fuel. The old man at first made the same excuse with those who had already declined acting as the traveller's guide; but perceiving him in great perplexity, and paying the respect due to his profession, 'You are a clergyman, sir?' he said. Mr. Walker assented. 'And I observe from your speech, that you are from the north?'-'You are right, my good friend,' was the reply. 'And may I ask if you have ever heard of a place called Dunnottar?'-'I ought to know something about it, my friend,' said Mr. Walker, 'since I have been several years the minister of the parish.'-'I am glad to hear it,' said the Dumfriesian, 'for one of my near relations lies buried there, and there is, I believe, a monument over his grave. I would give half of what I am aught, to know if it is still in existence.'-'He was one of those who perished in the Whig's Vault at the castle?' said the minister; 'for there are few southlanders besides lying in our churchyard, and none, I think, having monuments.'-'Even sae-even sae,' said the old Cameronian, for such was the farmer. He then laid down his spade, cast on his coat, and heartily offered to see the minister out of the moss, if he should lose the rest of the day's dargue."

Thomas Carlyle also commented on the dangers lurking along the moss road. In a letter to his mother in 1827 he noted:

"Had there been a moment of time left me last night, I would have written to you from Dumfries, and informed you four and twenty hours sooner of the happy issue of our tedious negotiation: but it was towards eleven at night before the higgling ceased and the papers were fairly signed; and then Alick and I had nineteen rough miles to ride, and moreover were afraid of being robbed by rascally Irishmen in the Trench of Lochar Moss, for we were carrying money in our pockets."

Given that the 'Trench' of the Lochar Moss is marked today by a cluster of service stations and motels on the A75 Euroroute, there are those who might claim that it is still a risk to those carrying money in their pockets!

Canals and Railways

Remarkably, the Lochar Moss very nearly got caught up in the 'canal-mania' of the eighteenth century. The entry for the Parish of Tinwald in the Old Statistical Account (1791) describes how *"about 30 years ago it had been proposed to build a canal from the Solway to Locharbridge at the head of Lochar Moss."* This appears not to be the 'new cut' of the Lochar, proposed primarily for flood defence reasons by The Society of Improvers and Smeaton, but a genuine transport link. Its route was from the Nith estuary near Caerlaverock Castle, by Lochar Moss, Torthorwald and Tinwald to join the Nith at Dalswinton. At that time, the canal system had proved very successful, bringing industry and prosperity to many areas, especially in England. Had such a canal been built, it could have provided an important outlet for agricultural produce and especially valuable for transporting sandstone quarried at Locharbriggs and in demand as building stone throughout the country. In 1812, Singer claims that the proposal had *"been lately in the view of a respectable civil engineer"* but it was abandoned due to lack of funds. Dumfriesshire sandstone would have to wait for the railways before it became widely used in the grand buildings of Glasgow, Edinburgh and other cities.

Constructed by the Glasgow, Dumfries and Carlisle Railway Company between 1846 and 1850, the railway line from Dumfries to Carlisle was authorised on 13th August 1846 with Dumfries Station and Racks Station opened in 1848. There was no village at Racks, the name meaning 'the old cart track down to the peats'. The station was built to serve various agricultural and peat-cutting businesses, and stimulated the growth of a village around it.

Figure 7. Racks Station *c.*1900

The rest of the line made no attempt to cross the Lochar Moss, instead skirting its eastern edge to the outskirts of Dumfries where the moss had already been reclaimed. It transferred to the Glasgow and South Western Railway in 1850, and the London Midland and Scottish Railway in 1923. Racks Station closed in 1965, but the line remains open providing views of Craigs, Racks and Ironhirst Mosses. With all three currently afforested, most modern passengers are unaware of their presence.

Exploitation: "an inexhaustible fond of moss"

Joan Blaeu's Atlas of Scotland of 1654 was the first attempt to publish maps for the whole country, but largely based on Timothy Pont's maps of the 1580s and 1590s, which still survive for Nithsdale. The Lochar Moss does not feature on either Pont's or Blaeu's maps, other than as area devoid of annotations. However, Blaeu included descriptions written by people other than himself of each area he mapped. The entry for Lochar Moss reads:

"At the second milestone from Dumfries is the famous peatmoss Lochar, 10 miles long, 3 wide: peats dug from here and hardened in the sun are burned by the whole neighbouring region. Lochar is divided in two by the Lochar Water, which floods in heavy rains and irrigates the surrounding fields with fertilising water; hence there is very great profit in hay."

Blaeu was Dutch and probably never visited the places in his atlas, but elaborated the maps and text. This suggests that in the sixteenth and seventeenth centuries Lochar Moss was a highly valued resource.

Early Uses

Blaeu refers to hay crops from surrounding fields, but it is very likely that 'bog hay' was also harvested from the Moss itself. Before agricultural improvements, this would have been in common use, not just from bogs, but from fens, marshes and other natural wet grasslands. Though it was considered to be of inferior quality, lacking in calcium and other essential minerals, it continued to be harvested even after the introduction of sown grass. Indeed, weather permitting, it is still harvested from a few sites in Dumfries and Galloway, though rarely from raised bogs.

Recent examination of the peat structure showed a curious phenomenon in the peat layers from at least one location and perhaps dating back to cultural activities almost 1,000 years ago. A double layer of heavily cracked peat lies beneath the current surface, the deeper one almost a metre down. Possibly attributed to modern forestry it is not typical of the known effects of such operations. However, it might result from ancient cultivation, most likely buckwheat. This is known from other parts of Europe but not previously from Britain and further investigation is needed.

Consumption of wild food is assumed, but not confirmed by documentary evidence until the '*New Statistical Account of the Parish of Mouswald*' (1845). By then, wild food had commercial value: "*in the peat bogs, the cranberry* Vaccinium oxycoccus *is gathered in large quantities, and sold in the Dumfries market.*"

Due to its absorbent and antiseptic qualities, *Sphagnum* has been collected and used to dress wounds for centuries, with evidence of use by the Romans on Hadrian's Wall. Perhaps Lochar *Sphagnum* was used by Roman soldiers in Ward Law. It was certainly collected from the Moss and other local bogs to make field dressings during World War One.

Another unusual use of the peat was to store butter. It is believed that butter was wrapped in skin or bark or placed in wooden kegs or wicker baskets and then placed in water-filled bog holes in spring and summer. This preserved it for winter use and mellowed the flavour. Probably dating from prehistoric times up until around the late eighteenth century, this practice was most common in Ireland and the Highlands and Islands. A sample of bog butter weighing 30gms, a very rare discovery in south west Scotland, was found in the private museum of Dr Grierson in Thornhill. Labelled 'Adipocere from the Locharmoss' and evidently cut from a larger sample, the date and details of its find are unknown.

Whether water was ever sourced from the Moss is unknown, but there is mention of Crichton's Well in the middle of the Moss.

The Moss was also a source of building materials. A small thatched cottage, built in the mid eighteenth century but of the type probably occupied by medieval peat diggers, still exists at Torthorwald. Its roof is supported on three crucks – pairs of curved oak trusses joined by a collar beam and pinned with pegs of ash. These retain the roundish section and irregularities of natural tree-trunks from which the bark was peeled. Two were not quite long enough for their purpose and needed additional extensions. Above the cruck trusses the roof was an arrangement of closely-spaced branch rafters, either of hazel or birch, over which there was a layer of turf providing a lining for the thatch. Now heather, in the recent past the thatch was straw. The original walls were probably clay, mud or turf; much originating from Lochar Moss. When the cottage was restored in the 1990s, the Moss was again the source of materials. In an area where quality timber was in short supply, it may be that the original crucks were from ancient trees unearthed from the peat. There are references to such timbers used by local carpenters. Lochar Moss peat certainly burnt in its fireplace.

Figure 8. Cruck cottage, still in existence at Torthorwald, and typical of the peat diggers' cottages.

Grazing

Grazing of bog vegetation with domestic livestock has probably been carried out for centuries, but as relatively poor-quality it was seldom recorded. Early writers often refer to black cattle as the stock put onto mosses, but sheep were also used. Grazing was probably restricted to the least productive animals, or else due to necessity like shortage of winter fodder or the need to remove stock from hayfields during summer. In 1743, The Society of Improvers noted that on Lochar Moss *"cattle only have Access to feed upon it in very dry seasons, (and indeed they do not chuse such grass except in the Months of April, May or June)"*.

Exact stocking densities are unknown, partly because stock may have ranged across the Moss been restricted to drier edges, but also because Moss grazing was only at certain times, with adjacent pastures used during other periods. Dr Singer (1812), discussing the poor management of bogs of Dumfries noted that winter grazing continued to late May giving bog vegetation little chance to recover. At all times, animals were watched over by herds to reduce risk of straying and 'bogging'. Losses occurred in ditches and through diseases like rot caused by excessive wetness.

Grazing became less common in the twentieth century, but a 2003 survey of farmers around Lochar Moss confirmed three farmers still grazed some cattle and sheep on the Moss but in conjunction with neighbouring fields.

Peat moss litter was bedding material for domestic livestock from the 1850s. It served to keep livestock clean and dry when kept indoors. *Sphagnum* litter was particularly useful as it absorbed large amounts of liquid and neutralised odours. Used mostly by local farms, there is little evidence of large-scale litter extracted from Lochar Moss, as occurred on other raised bogs.

Domestic Peat-cutting

If the 'impassableness' of the Lochar Moss was a key reason for the founding of Dumfries, the value of its peat was just as important to the town's population growth during medieval and early post-medieval times. Alexander Fenton in his classic study of Scottish rural life (1976) noted how eighteenth-century areas like Tongland in Kirkcudbrightshire lacked peat resources, so *"there was real distress for fuel, which many poor and shivering wretches suffer."* In despair, alternatives like animal dung were burned. By contrast, Lochar Moss peat was considered inexhaustible with domestic fuel cutting the main use of the Moss until the twentieth century.

One Irish calculation reckoned that, in the absence of other fuels, about three tons of air-dried peat were required per person per year, equating to the digging of around 24 tons of wet peat. Another calculation for a Scottish peatland one mile from a road and three and a half miles from the fireplace, estimated that to cut 16 tons of peat, dry it and transport it, required 15 days of work (12 hours per day) for four people and the work of two horses for 14 days. Of this, one man could complete the digging in three days; the rest was required to transport it home. The peat wealth of Lochar Moss therefore required considerable effort, but on the doorstep of Dumfries and with few alternatives, it was of enormous value. Wood had all but gone from around the town well before the medieval period. In any case, wood had alternative uses, required even more labour to harvest, was more difficult to transport, and did not provide the same heat output ton for ton. Mineral coal was not available locally, having to be brought in from Sanquhar or Byreburn (Canonbie), or from Cumberland. Also, until the end of the eighteenth century there was a high tax on imported coal; even after this date it was subject to custom fees. Samuel Lewis, in his 1846 parish account of Mouswald in '*A Topographical Dictionary of Scotland*' described the problem:

"*Peats are the fuel commonly used. The Duke's tenants get theirs from the moss within the parish. The other proprietors tenants get theirs from the same moss in the parishes of Torthorwald and Ruthwell; and, though there is an inexhaustible fond of moss, and they have peats for the casting, winning, and leading, yet they consume a great deal of time, which might and would be better employed to much better purpose in the management of their farms, were coals to be got at a moderate distance.*"

Steele (1826) in '*The Natural and Agricultural History of Peat-moss or Turf-bog*' primarily a discussion on the best methods of draining and reclaiming bogs, conceded the importance of peat in areas without ready supplies of coal: "*The people of Holland (says Anthony Von Walter), may thank God that they have so many peat-mosses, which are worth mines of gold to them.*"

Probably because of its high value, from medieval times peat cutting was a highly organised and regulated activity. One of the oldest documents in the Dumfries Archive Centre is a Warrant from James V, dated 1524, ordering his sheriffs to settle a dispute over Lochar Moss. It had been used as common land for digging peats by the inhabitants of Dumfries until claims of possession were made by Thomas Gledstanes of Kelwood. At that time, those caught stealing had their cheeks branded with the Dumfries burgh clock key heated on a fire made from stolen peats.

Three hundred years later, peat-cutting was still the subject of legal cases. In 1812, the Marquis of Queensberry took three of his tenants on Craigs Moss to court to prevent their cutting peats for sale, or letting out peat cutting to others. The defendants claimed that residents of Dumfries, Maxwelltown, and adjoining villages had always received their supply of peats from the moss. By usage from time immemorial, a privilege of winning peats for sale had been annexed to several of the farms on Craigs Estate and even allowed in many instances on several farms some distance from it. Though they lost their case, the Marquis allowed them to continue cutting peats for their own domestic use.

Regulation of Moss peat cutting increased into the nineteenth century. A document of 24th May 1845 entitled 'Regulations to be observed by all those that cut peats in the Craigs Moss' stipulates that each area of the Moss leased to the various farms should be clearly marked. Cast peats were to be neatly laid in such a manner as to avoid getting mixed up, and that nine pence was to be paid to John McDougal of Barbeth *"for each yard in the breadth of their lease"* for repair work. The document was signed at New Abbey, with Barbeth and other named farms in the New Abbey area. Transporting cast peats across the River Nith must have been a regular sight. There would be similar agreements for peat cutting rights in Dumfries and the mosses on that side of the river.

Peat-cutting was carried out mostly by men. The spade was the standard tool, but different regions developed their own distinctive local implements. The example below is from nineteenth-century Wigtownshire, but similar implements were used at Lochar Moss. It has a sharp metal blade with one side turned up at right-angles to form a sloping wing for cutting brick-shaped peats from a vertical face, and a flattened section of the wooden haft to support the peat and enable it to slip off easily onto the stack.

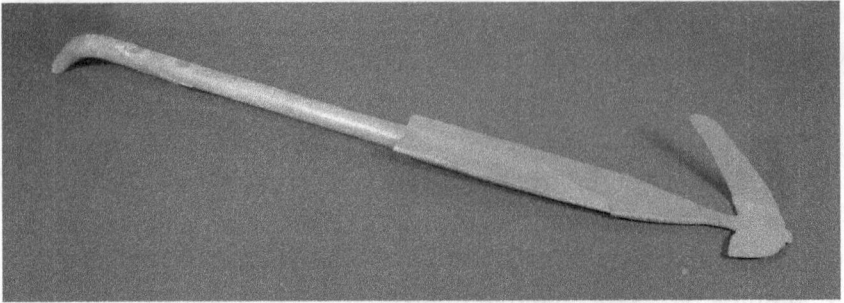

Figure 9. Peat spade from Wigtownshire © Dumfries & Galloway Museums

Women and children were also involved, probably in stacking peats to dry, turning them whilst drying, and transporting them offsite by sledge, barrow, cart, or pannier. The latter was either entirely human-powered or assisted by donkeys or ponies. A curious contraption known as a bog tramper or horse patten is known to have been fitted on the hind feet or on all four hoofs of horses ploughing reclaimed bogs in Wigtownshire. In 1812, William Singer noted *"horses with wooden clogs on their hind feet" on the Lochar Moss.* These prevented 'boggings' when a horse sank into the soft ground and had to be dug out. Trampers could put an undue strain on a horse's tendons and many animals were lamed by wearing them.

Figure 10. Bog tramper or horse patten, Wigtownshire © Dumfries & Galloway Museums

In his *General View of the Agriculture of the County of Dumfries* (1794), Bryce Johnston, Minister for Holywood, quotes local labour rates for peat cutting; men were paid one shilling and threepence for a 12-hour day, women 'working at peats' got only ninepence.

Those who worked cutting peats on the moss were seldom referred to positively. Dr Johnston's journal of a short visit to Jardine Hall in 1844 is typical:

"Lochar Moss supplies the good people of Dumfries with an abundance of peat, which is the fuel with the commonality all over this district, and there were workers of it scattered throughout the moss. There is a certain interest about these men, who appeared to be of the lowest class in general. No noise attends their monotonous labour, the spade cuts without grating, the clod is thrown aside without evoking a sound, there is no converse, each toils by himself, without giving or receiving another's orders or directions; silence reigns around, and imparts to the labour a peculiar, but rather disagreeable, interest; for this outward solemnity of nature tells not favourably on the minds of men of the low degree of cultivation these have. Solitude is not for them."

Figure 11. A Galloway Peat Moss by William Stewart MacGeorge (1861-1931) shows a typical peat-digging scene of late nineteenth or early twentieth century.

Though the main area for domestic peat cutting appears to have been in Lochar Moss closest to Dumfries, in areas subsequently drained and either built on or converted to farmland, some surviving peatland shows signs of extensive domestic peat cutting. This includes large areas of Craigs Moss, an area of Racks Moss south of the village, and a substantial part of the western half of Longbridge Muir, with a smaller area along its southern edge. It is possible that some of the peat extracted from the southerly locations may, in addition to fuelling domestic fires, have been used to burn seaweed to produce kelp. This material was used in the soap and textile industries. This small-scale industry was practised on the Ruthwell shore in the late eighteenth century, with records of kelp export to England. Ironhirst Moss was least subject to domestic cutting, but that made it more attractive to subsequent commercial peat extraction.

Agriculture

Despite the East Anglian Fens being drained from the mid-seventeenth century, there appears to have been little serious attempt at agricultural reclamation of Lochar Moss at that time. Only at the southern end of the Moss, where the natural processes produced agriculturally valuable and workable soils similar to those in the English Fens, was agricultural reclamation completed at an early date. Any peat formed here was likely to be thin and easy to plough into the underlying sediments to create workable land and leaving little trace of peat. Elsewhere, the Moss remained largely intact until the mid-eighteenth century.

By 1743, The Society of Improvers in the Knowledge of Agriculture in Scotland highlighted the agricultural potential: *"and it is certain, that the greatest part of this very Moss of Locher, if once sufficiently drained, might, on a very reasonable expense, be made as profitable ground as perhaps any that can be found in either of the Counties of Nithsdale or Annandale"*. They knew little of how the Moss had formed or the natural processes maintaining it, but they worked out the correct sequence of events: *"this valley, which appears to have been once Sea, and thereafter a Wood"*. The suggested action to drain it involved removing Lord Maxwell's mill at Bankend which acted as a dam on the Lochar Water; digging a canal to take water from the Mouswald Burn and prevent it entering the Lochar Water; widening and deepening of the channel of the Lochar itself; and the digging of extensive drainage ditches, including one encircling the whole Moss. It was deemed especially important *"to make as many principal drains, quite through the Mosses, as may be necessary, of a sufficient wideness, and of a deepness below the very bottom of the Moss: For, until every drop of the very heart's blood, as it were, be let out of it, it will never give over growing"*. All this speeded water-flow through the Moss with an added bonus of the mill relocating to the new canal, which provided more constant water-flow. The scheme's main proposer, Mr Maxwell, advised the Duke of Queensberry and Dover, the main landowner, that *"your Grace, on very moderate charges...engage a Servant sufficiently qualified to carry on the Works, (upon which the success of the Design depends)"*.

The man appointed as 'the sufficiently qualified servant' was engineer John Smeaton, famous for designing buildings, bridges, canals and other structures such as Eddystone Lighthouse. He designed several well-known structures in Dumfries and Galloway, such as Dumfries Town Mill (now the Burns Centre) and Portpatrick harbour, but his first civil engineering consultancy was to devise a scheme to implement the drainage. His list of Dimensions and Estimates, dated 21st September 1754 and totalling £2,952, was prepared for Charles, Third Duke of Queensberry. The main items were a large sluice gate near the Solway to stop the lower reaches of the Lochar Water being tidal; straightening, widening and deepening the channel of the Lochar Water between Upper Locharwoods and Caerlaverock Church (3 miles), Mouswald Burn and Tinwald Isle (9 miles) and Tinwald Isle to Locharbriggs (1½ miles); and a new mill on the "new cut" on the Mouswald Burn. His estimate still exists, archived in Dumfries, and attached to detailed observations relating to flooding. The latter are dated eight years after the estimate. His sketch-plan also exists, though now detached from the estimate and archived with the Royal Society in London. Smeaton clearly states that the main benefits to his proposals would be to farming: *"The advantage immediately arising from this would be a complete drainage of the meadows upon Lochar all the way from Locharbridge to the Isle House and be complete safe-guard to the crops."* Despite the document title referring to drainage of Lochar Moss, he did not claim the proposals would completely drain all the Moss and allow farming of the peatlands, but *"the draining of the mosses in general will be rendered so much more easy and practicable...Even Lord Stormont's Moss* [Cockpool Moss] *which seems to be most remote from the benefit of this work may receive some advantage from it."*

Successful implementation of Smeaton's plans was hindered by the requirement to have agreement, and no doubt financial backing, from a number of different proprietors. Several subsequent observers stated that they were never implemented. No doubt the continued existence of the Lochar Moss into the nineteenth and twentieth centuries encouraged such a belief. However, given that it was not Smeaton's intention to completely drain the whole Moss, it should not be assumed that his plans came to nothing. At the northern end of the Moss, a 'New Cut' of the Lochar Water between Tinwald Isle and Locharbriggs was dug as recommended by Smeaton and can be followed today. The modern Ordnance Survey clearly shows both the 'Old Course' and 'New Cut'. Indeed, the 'New Cut' is rather more extensive than suggested on Smeaton' plan. William Crawford's map suggests it was not there in 1804, but this is perhaps a mapping error as a 1799 Plan of the Farm of New Mains of Tinwald (now held in the National Archives of Scotland), though not extending as far as the 'New Cut', does refer to 'The Water of Old Lochar'. Both channels are clearly marked on the first edition Ordnance Survey map in 1855. At the south end of the Moss, there is less evidence of large-scale improvement works but a 1776 estate plan of Bankend shows the mill, not on the main Lochar Water channel, but its own specially constructed mill lade. This channel can be traced today, though the surviving mill buildings are a short distance from it.

During the latter years of the eighteenth century, when the national fervour for agricultural enclosure and improvement peaked, considerable effort was put into other measures to reclaim the Moss for agriculture. Extensive drainage ditches were constructed and peat was consolidated by application of sand, clay and lime. William Singer gives a detailed description in his *'General View of the Agriculture of Dumfries'* (1812) of *"the remarkable improvements executed by the late Baillie Shaw...on portions of Lochar-moss...which he has improved into fields, with hedge-rows and crops of all sorts, not excepting fruit trees."* For initial applications of sand, sea-sand from the base of the peat was recommended: *"Where sleetch is to be had in full quantity, no top-dressing answers better on peat-moss."* 'Paring and burning' was another recommended technique which involved cutting and burning the peat surface layers, then mixing the ashes with peat in the new surface layer. It is possible the burning got out of hand on occasions a major fire reported in 1785, then others in 1826 and the 1930s. The latter was probably caused by sparks from a train. Some techniques proved successful, the 1799 Plan of the Farm of New Mains of Tinwald showing arable in crop or fallow, pastures, lawns, orchards, plantations and field boundaries. Not all this farm was originally peatland, but other than a reference to 'moss pasture', there is now little indication of surviving peat.

However, even in the early nineteenth century, not everyone welcomed drainage of mosses and other wetlands. James Grahame, the poet, lawyer and in later life, curate, lamented the effect of agricultural improvements on birdlife. His wife hailed from Annan, and in 1808 he composed 'British Georgics' (poems dealing with agriculture or rural life) whilst spending the summer at Mount Annan Estate, rather ironically owned by one of the leading 'improvers' in Dumfriesshire. In this, he expressed his opinion on the effects of drainage and with reference to bitterns (at that time restricted in Scotland to the Lochar Moss), probably had the Moss in mind:

"No more the heath fowl there her nestling brood
Fosters; no more the dreary plover plains;

And when, from frozen regions of the pole,
The wintry bittern to his wonted haunt,
On weary wing, returns, he finds the marsh
Into a joyless stubble ridge transformed,
And mounts again to seek some watery wild."

Concerns such as those expressed by Grahame had little impact and in 1827 large-scale proposals for draining Lochar Moss were resurrected. A document entitled 'State of facts relative to the water of Lochar and Moss thereof' largely repeats, word-for-word, Smeaton's observations. This document was probably by Walter Newall, who prepared plans of the area which still exist. Once again there is little evidence of works being followed through.

In the 1830s, John Heathcoat, MP for Tiverton in Devon, offered an alternative course of action through his invention, a steam-driven plough. News of his machine was greeted enthusiastically by farmers, for it offered not only the prospect of ploughing faster and cheaper than with horses, but also the hope that land previously unavailable for cultivation could be brought into production. A meeting of the local committee of the Highland and Agricultural Society of Scotland agreed to bring the plough to Scotland so that it could be seen in action as part of the Society's forthcoming show in Dumfries. The total cost of the entire show, including the construction of a purpose-built pavilion capable of seating 1,000 for dinner, was estimated at 400 pounds. Some idea of the perceived importance of the steam plough can be gauged by the fact that 250 pounds of this budget was allocated to transport the plough by sea from Lancashire, where it had been undergoing trials near Bolton, and to demonstrate it in action on Lochar Moss.

Specially designed to slice through heather and not to get entangled in *Sphagnum* moss, the plough was made to cut a furrow of 23cm (9 inches) in depth, turning the cut earth as it went. Though weighing 25 tons, 30 tons when fully laden with fuel, the complete machine was mounted on 12 wheels, each 2.5m (8 feet) in diameter, which moved on *"an endless flexible railway"*, thereby giving a ground pressure only *"4¾ times more...than a man does while walking."* The inventor therefore claimed that it could operate in locations *"where a cart would be swamped."*

Arrangements were made and the house-sized plough duly arrived at Glencaple Quay. It was moved three miles overland, at the substantial cost of 15 pounds, to a part of the Racks Moss farmed by Mr Paterson of Kelwood, who paid 50 pounds for the privilege. It was set up by Josiah Parkes, engineer to Heathcoat, and was to be operated by *"three or four stout Englishmen trained to the business."*

Figure 12. Illustration of Heathcoat's Steam plough from Farmers Weekly 1837

When the first day of the show arrived on 4th October 1837, interest in the plough exceeded all expectations. The fascination was not confined to *"the local districts immediately connected with it, but in the other portions of the kingdom, and even abroad. Distinguished visitors from England, Ireland, France, and Germany, combined to render the meeting one of national importance."* All went well on the first day of ploughing. Invited guests and others paying a shilling witnessed the successful ploughing of 8 acres in 12 hours. Indeed, the Dumfries Courier reported that *"we may expect to see, at no very distant period, that immense tract of barren morass – now only the resort of curlews and seagulls – become a fertile valley, covered with clover and wheat."*

The second day also went well, but problems began on the third morning, described by Parkes as *"that dismal morning"*. Heavy rain made conditions difficult, but this did not deter the crowds with an estimated 2,000 visitors taking up the offer of viewing the spectacle for sixpence. This led to the bog becoming *"poached and mashed"*. Not only that, but the unmanageable spectators swarmed all over the machine, further restricting Parkes' chances of demonstrating it in action. The result was that very little ploughing was carried out that day and most spectators left disgruntled and unimpressed with the steam-plough's capabilities. Following a vote of thanks to Mr Heathcoat at the show's dinner, the otherwise very detailed report of the show published by the Society records only that *"Mr Heathcoat returned thanks; but we regret to state, that from the position in which we were placed, and the low tone in which he spoke, we were unable to catch the purport of his observations."* His state of mind can only be surmised.

Parkes tried to salvage the situation, writing to the *Dumfriesshire and Galloway Herald* that *"Mr. Heathcoat is ready and desirous to undertake the reclamation of any eligible and extensive tract of bog; and that he and myself are ready to treat with any proprietor of bog disposed to employ us."* But the damage was done, with one commentator describing Lochar Moss as the cradle and grave of the first steam plough in Scotland. *The Farmers Magazine* (1840) described how *"his plough apparatus still remain unemployed on Lochar Moss, exactly as it was laid up after the great meeting in Dumfries"* and another commentator, 30 years later, described how *"both engine and plough sank"*. Around 1890, some parts of the steam plough were salvaged, but much may still lie buried somewhere on Racks Moss. A 1962 expedition to find it was unsuccessful. Perhaps as according to some reports, all the useable parts were earlier salvaged, exported and put to use in Egypt.

Steam-ploughs did provide useful service on the drier lands of eastern Scotland, but failed to play any significant part in the reclamation of peatlands. Given that Paterson, owner of the part of the Lochar Moss where the trials took place, was a descendent of William Paterson, the leading promoter of the ill-fated Darien Scheme that almost bankrupted Scotland, the loss of his 50 pounds in Lochar Moss was far from the worst of family investments.

Reclamation of the Lochar Moss did, however, continue without the aid of steam ploughs. *'The Topographical Dictionary of Scotland'* (1846) *"portions of the Lochar moss were some years ago brought into cultivation, yielding abundant crops of oats, potatoes, and rye-grass"* and that the *"master of the school at Collin has a salary of £20, with a house, and three-quarters of an acre of land reclaimed from the moss."*

In 1878, detailed 'Plans for Improvement to River Lochar' were prepared and still exist. It was essentially a return, for the third or fourth time, to schemes formulated by the Society of Improvers and John Smeaton over 120 years earlier. Though the idea of a canal and tidal sluice-gate was dropped, the plans proposed locations for straightened and deepened sections of Lochar Water with associated drains. Again, implementation was limited.

One unusual venture in agricultural reclamation was undertaken by the Scottish Labour Colony Association Limited at Mid Locharwoods from 1898 to 1917. This Association was set up to provide food and shelter to unemployed men in exchange for work. Part of this work was *"drainage of the moss land, its reclamation by liming and claying, and its ultimate conversion into arable land, capable of growing crops of timothy, cabbages, carrots, and potatoes."* Peat was cut from the Moss with the Association planning to expand this activity: *"The moss land lies at one extremity of Lochar-moss, which covers some thousands of acres, and there exist, therefore, possibilities of a great expansion of the operations of the colony."*

By 1919, large-scale drainage plans remained subject to discussion rather than implementation. Sir Francis Fox, a civil engineer involved in designing Sydney Harbour Bridge, the

Mersey Railway Tunnel, and the peat factory established on Ironhirst Moss (see below), wrote to Sir James Crichton-Browne, leading psychiatrist at The Crichton Royal Institution in Dumfries. He advised that *"The people in Dumfries and Lochar Moss should form a committee and raise the necessary funds for the preparation of plans, estimates and report."* There is no evidence of this happening. Small-scale drainage schemes for agricultural purposes continued. A Lochar Moss Drainage Committee formed under an Improvement Order through the Land Drainage (Scotland) Act 1958 to be responsible for deepening Lochar Water to improve land drainage. Though the actions of the Committee succeeded in lowering the water level adjacent to the mosses, since the late twentieth century some farms entered their moss-lands into agri-environment schemes and are paid to manage them to benefit the raised bog habitat.

Commercial Peat Extraction

Despite a long history of domestic peat-cutting and agricultural reclamation, the vast supplies of peat remaining at Lochar Moss at the end of the nineteenth century must have appeared as fruit ripe for picking. With minimal investment in men and machinery, endless supplies of peat could, it seemed, be extracted from the 'wasteland' and sold, at great profit, for fuel, power or other purposes. However, despite repeated attempts during the twentieth century, both by private entrepreneurs and public bodies, the promised fruits never materialised.

In the nineteenth century, there were attempts at commercially selling hand dug peats. The Peat Supply Company operated from Blackgrain on Longbridge Muir in 1890. But early in the twentieth century, the first machine peat plant was put into operation near Racks Station, aiming to mass produce peats for the domestic market. William Kerr, commenting on the project in his 1905 'Illustrated Treatise on Peat and Its Products as a National Source of Wealth' was enthusiastic and optimistic about the future:

"In Dumfriesshire, on the extensive Lochar bog at Racks, the Scottish Peat Industries, Limited, under the able superintendence of Mr. A. B. Lennox, have established an extensive and rapidly extending factory, which, with the vast amount and excellence of raw material, and the facilities of getting the manufactured fuel and other various bye-products, etc., on to the rail and to a port, promises to prove a source of profit to its shareholders, and a reliable and interesting object lesson of strenuous labour and intelligent enterprise."

The plant appears to have had only a very short life.

Figure 13. Scottish Peat Industries Limited at Racks Moss *c*.1900

Meanwhile, on nearby Ironhirst Moss in around 1910, Peco Ltd. put into operation methods of gasifying the peat to produce sulphate of ammonia for the chemical industry. This too proved a short-lived operation when more economical methods of production were introduced elsewhere in Europe. Instead, the company switched to the wet-carbonising process to produce peat for fuel, and following the outbreak of the First World War the Government stepped in to build a wet-carbonising plant at Ironhirst that was designed to produce a large quantity of peat briquettes of very good quality for use by troops in the trenches. The peat was harvested by a floating dredger and pumped to the plant through a pipeline. An area of approximately 20 acres (8ha) was excavated in this way, but the plant was not completed in time for large supplies to be made available for the War effort. Keeping water levels low enough to allow extraction also proved difficult, and this area is now a large water-body known as Ironhirst Loch. The plant was closed soon after the War and partly demolished.

Figures 14 and 15. Ironhirst Loch, created by peat extraction, and the remains of Ironhirst peat factory © Peter Norman

Peco again took over the factory site together with Racks and Ironhirst Mosses, and continued to carry out dewatering experiments and other research. The main result was the development of the Peco drier and the Peco milled peat method. This method first required the cutting of parallel open drains over the bog surface using a plough ditcher. These were approximately 15m (50 feet) apart, up to 1.4m (4 feet 6 inches) deep and about 1.5m (5 feet) wide at the top and effectively, split the whole working surface of the bog into a grid of fields. The drains discharged at either end into piped outfalls which ran at 90 degrees to the open drains and which themselves discharged into the nearest natural outfall.

The peat was worked by the specially designed Peco disc ditcher, consisting of a cutting disc rotating at a speed of 124 revolutions per minute carried on an arm offset from a tractor. Working speeds varied from 70 to 650 yards per hour depending on working conditions, and though the complete machine weighed more than 17 tons, timber swamp shoes and front rollers reduced its ground pressure, ensuring that it did not befall the same fate as Heathcoat's steam plough. Beyond the piped outfall, a headland or turning ground 20-30m (70-100 feet) wide was left to enable production machines to turn from one field into another. A permanent narrow-gauge railway was laid to transport the milled peat from the bog to the factory where it was subject to indirect drying by means of hot water and steam. It is likely that there were temporary rail spurs to the fields currently in production.

Between the Wars, optimism remained high that a commercial success could be made of this method of extraction and Peco produced plans for a briquetting plant to be built at Ironhirst to use all available peat from Ironhirst, Racks, Longbridge Muir and Craigs Mosses. The *Colliery Guardian and Journal of the Coal and Iron Trades* carried the following note in its edition of 20th July 1934:

"It is hoped shortly to arrange the necessary finance to carry out the enlargement and reorganisation of the factory at Ironhirst, near Dumfries, belonging to Dumfries Peco, Ltd., in which Peco holds 98 per cent of the issued share capital. In the meantime work is proceeding on the bog at Ironhirst for the production this season of the necessary supply of raw material for use in the factory when the same is ready."

However, competition from coal proved too great and during the Second World War Peco ceased operations at Ironhirst. The Peco method of producing milled peat was the standard technique on many bogs in Britain and Ireland up to the present day.

Following two seasons of hand-cutting by German Prisoners of War, post-war machine peat production on the Lochar Moss was restricted to a single season. However, the Dumfriesshire Peat Fuel Scheme was an experiment to test whether production of air-dried peat by hand-cutting methods could be economically developed to contribute towards fuel problems. It was soon added to the ever-lengthening list of short-lived peat extraction ventures on the Moss. Hansard, the official published report of proceedings in Parliament, included a written question from The Earl of Mansfield in the House of Lords, himself a landowner of part of the Lochar Moss. On 26th May 1949 he asked His Majesty's Government, regarding the 'Dumfriesshire Peat Fuel Scheme':

"….whether they will state the total expenditure incurred on the now abandoned Ministry of Fuel scheme for producing fuel from the Lochar Moss, Dumfriesshire, together with the tonnage of fuel extracted, and the final loss to the taxpayer."

Lord Macdonald of Gwaenysgor responded:

"The total expenditure on the operations conducted by the Ministry of Fuel and Power at Lochar Moss, Dumfriesshire, including the net cost of equipment, amounted approximately to £9,500. The scheme produced about 2,600 tons of saleable air-dried peat, which was sold for approximately £8,600. The net loss was therefore £900."

It seems another failure was not enough to dampen belief in peat as a vast untapped resource. In 1949, the Secretary of State for Scotland set up The Scottish Peat Committee to assess how the country's substantial peat resources could be best utilised. They published their first report in 1954, identifying lack of accurate survey information as a significant problem preventing exploitation. They recommended peat surveys be continued until all major Scottish peat deposits were examined and classified.

The Department of Agriculture and Fisheries for Scotland, under the guidance of the Committee, carried out a survey of Lochar Moss in 1957. They published the results in a report to the Secretary of State in 1964, with the expectation that *"the information gathered together will...enable future users of peat to assess the possibilities for practical exploitation of this vast natural resource."* It was described by the Committee as *"the most suitable peat deposit for exploitation in the south of Scotland and perhaps even the whole of Scotland"*. They identified few obstacles to exploitation. If necessary, Ironhirst Loch could be easily drained. They noted that *"only one item of unusual interest has ever been found"*, the bog body discovered in 1947, not the Iron Age torc in the British Museum nor any other archaeological remains. Government optimism was supported by others in the scientific community. *New Scientist* (13th May 1958), featured 'Peat Winning: An Ancient Industry Transformed' with "*peat remains important as a fuel, in spite of the advent of nuclear power; it requires only a relatively small amount of capital for its exploitation.*"

Despite this survey and optimism for commercial extraction of Lochar Moss peat, from the mid-1960s attention turned to an alternative use of the Moss, afforestation. If this could not produce direct income from the peat itself, it would at least produce saleable crops from the land. That did not mean all hope of extracting Moss peat was extinguished. Even with most of Lochar Moss planted with trees, attempts continued to extract commercial peat. This time it was not driven by private companies keen to move onto the site, but by the Forestry Commission and a Regional Council, keen to persuade them to do so.

In January 1981, Racks, Ironrhirst, Longbridge Muir and Holmhead Mosses, a total of 1,000 hectares with an estimated 40 million cubic metres of peat, were offered for peat working on a 25-year lease on condition that standing timber should be allowed to grow until it could be harvested at its normal age. The Forestry Commission planned to replant after peat removal. Only Ironhirst Loch and the adjacent area were excluded for 'conservation / amenity' reasons; rather ironic given that this loch was the one part of Lochar Moss entirely created by previous commercial peat extraction. In January 1982, the offer of a lease changed to an offer to sell, complete with published prospectus, and accompanied by a Council publicity leaflet entitled 'Growing in Peat'. The statutory conservation bodies had few concerns, only requesting that a few well-distributed ponds were left after extraction.

A number of companies (including some based in Ireland, Finland, Belgium and The Netherlands) expressed interest in peat extraction for both horticultural and energy use. The investment required was quoted at between £6 million and £60 million resulting, it was claimed, in 70 to 250 new jobs. On 16th January 1986, the Regional Council's Industrial Development Officer wrote to his Chief Executive: "*After many disappointing false starts, it would appear that we now have a company with unlimited cash resources who are determined to fully develop this natural resource.*" But it was not to be. The company withdrew, followed, one-by-one, by all the rest, quoting the high set-up costs required and uncertain markets for peat products. Perhaps some were never that serious. A fact-finding visit to Scottish peatlands by two Dutch business executives very nearly ended before it had started, as they had not realised that wellington boots would be required to visit a bog! Two complimentary pairs were hastily acquired from the Dumfries factory where they were made and the Lochar Moss visit went ahead. Their subsequent letter to the Council began by commenting on how comfortable the wellingtons were, and how useful they proved to be in the Netherlands. However, it was less than enthusiastic about a commercial venture on the Moss.

Interestingly, the last company to withdraw, Fisons in 1990, quoted the environmental lobby as their main reason and "*public protest which it is feared would arise were Lochar Moss to be developed*". Evidence of such a lobby came in the form of a letter received by the Council from Friends of the Earth the same year. This referred to the Council's 1981 'Growing in Peat' leaflet and argued that it ignored the value of the Moss as wildlife habitat and its role in water regulation and carbon sequestration. The response from the Director of Economic Development showed just how much things had changed in a short time. Apart from pointing out that the leaflet was out of date, he conceded that the future of the peat at Lochar Moss was likely to involve conservation, rather than exploitation. In a letter to the Forestry Commission in November 1990, he confirmed that "*the Council has no plans to try to attract commercial peat interests to Lochar Moss.*"

Forestry

It seems likely that a few trees have always grown on Lochar Moss. Roy's Military Survey map of *c.*1755 shows an enclosed square near Comlongan, clearly surrounded by bog. Within this enclosure, his symbols are not the ones he usually used for trees, but perhaps he meant to indicate young plantation. Certainly, early nineteenth-century maps show enclosed trees at a similar location. They are not marked on 1850s first edition Ordnance Survey maps, but these show a small clump on Holmhead Moss, a few on the north of Ironhirst Moss, and the greatest concentration on Carnsalloch Moss, marked as 'Burnt Firs' but now part of Heathhall Forest.

All the above trees were planted, with the exception of those at Holmhead Moss that appear natural. Today there is a narrow belt of Alder, Willow and Birch here, hemmed in by deep drainage ditches. It is possible these ditches prevented access to this area during conifer planting and the trees subsequently colonised unplanted areas. More likely is that these are the trees shown on the 1850 map. The Alders appear to have been coppiced, cut at ground level to encourage vigorous regrowth harvested on a regular cycle. During the eighteenth and nineteenth centuries, this was a common practice and few trees escaped, even in remote locations. But the regrowth from the Alders at Holmhead now emerges from a point at least 0.5m (2 feet) above current ground level, with former roots now exposed and supporting the rest of the tree on strut-like aerial roots. It appears that the last time these trees were coppiced, the peat was deeper. Occupying an area between the raised domes of Ironhirst and Longbridge Muir, they are natural survivors from a time when most of the rest of Lochar Moss was treeless. It is normal for peatland habitats in such locations, a habitat which scientists have termed a lagg fen, to be different to those in the centre of the dome.

From 1967, the Forestry Commission acquired large areas of Lochar Moss and began ploughing and planting. The western half of Longbridge Muir, subsequently known as Black Grain

Plantation, and Cockpool Moss were planted in 1967-8 and Racks Moss in 1969. By 1979, all four surviving peat domes were transformed by commercial afforestation; mostly of Lodgepole Pine with some Sitka Spruce. Only Longbridge Muir retained substantial raised bog habitat, with a smaller area on Racks Moss. This resulted in the widespread changes in surface hydrology and loss of light. However, the crucial difference between this and drainage for agriculture and commercial peat extraction is that forestry management does not destroy the dome of peat, the fundamental feature of all raised bogs. Any species that can survive the physical disruption caused by ploughing and the competition for light can still benefit from a relatively high water-table. So it was at Lochar Moss.

Though the plantations resulted in substantial losses, to the point of extinction for some specialist bog plants and insects, they could not eradicate the bog vegetation entirely. Consequently, even 35 years after the planting, it was still possible to find typical raised bog vegetation, in a somewhat reduced form, hanging on under dense mature plantations. When, in the early years of the twenty-first century, areas began to be felled, re-ploughed and re-planted, remnants of *Sphagnum* had opportunity for resurgence, spreading slowly and imperceptibly as cushions, hummocks and carpets across the bare peat surfaces. Even large amounts of brash, mechanically-stripped from conifer trunks and left lying on the ground, could not check the march of these small water-dependent plants. On the contrary, brash and even large logs were smothered in fresh mossy growth, furrows were choked in *Sphagnum* within a year of ploughing, and small pools formed where drains had been dug to remove surface water.

All of this was of course only a temporary resurgence of the bog. Though the *Sphagnum* and other bog vegetation were able to reassert their presence for a while, the main drains continued to operate efficiently and remove water from the site and Racks Moss was replanted in 1999-2000. Nor will the peat hang on indefinitely. As tree roots take up so much water, the water-table gradually falls and as the surface dries, deep cracks develop in the upper layers (1-2m) of the peat. Two sorts of cracking have already been identified. A series of deep vertical cracks developed beneath drains and plough furrows, and a layer of dried peat grew a net-like pattern of cracks. This cracking does not necessarily lead to water losses from the peat, but if drains are cut through then these breach the cracks causing them to drain the site. In many places along the sides of freshly-dug drains on all three of the raised bogs surveyed in 2003, was evidence of cracks breached and causing outflows. In some places water was seeping slowly, in others it flowed freely.

Changing Land-use

Although undated, the Dumfriesshire maps produced as part of General Roy's Military Survey of Scotland probably relate to 1752-55. Only the main landscape features were noted by Roy's small team of engineers. Towns, settlements, enclosures, woodland, and relief were all sketched in by eye or copied from other maps. Consequently, it is not possible to take accurate measurements from Roy's maps, but considering the difficulties of access and surveying such an environment, his depiction of drainage patterns east and south of Dumfries is remarkably accurate. Except for the road from Dumfries to Rockhall, Lochar Moss is shown as a vast unclaimed wilderness stretching unbroken all the way from Locharbriggs to the Solway. At this time, expansion of Dumfries may have already nibbled at the edge of the Moss, but the sinuous boundary suggests that this had not occurred elsewhere. Only three small settlements or farms (Upper, Mid and Lower Dargavel) are shown within the Moss. Roy's map is therefore the best depiction we have of the full extent of Lochar Moss.

A hundred years later, following a period of intense agricultural change, the Ordnance Survey produced their first maps of the area. Using these maps, in 1984 the Nature Conservancy Council made an historical assessment of the extent of Lochar Moss. In 1855/56 the combined area of the Lochar Mosses was estimated to be 2,311 hectares. Clearly peat cutting, drainage, and agricultural reclamation had already impacted on the peatlands, but all the areas mapped by Ordnance Survey appeared dominated by undrained mosslands. By 1899, the total area had reduced only slightly to 2,281ha, but 8% of the area was farmland and 3% under woodland. However, by 1973/74, out of a total area of 2,231ha, only 15% remained as moss. The rest had been highly modified, by far the largest proportion (36%) being under forestry, 8% under other woodland, 13% farmed, 19% drained moss, 9% peat cuttings and a tiny percentage had been built upon.

Wildlife: "particularly favourable to the growth of Cryptogamic plants" - Mosses and Liverworts

James Cruickshank's statement about Lochar Moss being *"particularly favourable to the growth of Cryptogamic plants"* (i.e. non-flowering species reproducing by spores rather than seeds) is hardly surprising given the critical role of mosses in the creation of peatlands. In his 1842 'List of Jungermannia &c. observed in the neighbourhood of Dumfries', he lists a wide range of liverworts from Lochar Moss. In a style unlikely to be seen in modern scientific reports he describes the extent of *Jungermannia Lyellii* as *"the whole patch might be covered by a man's hat"*. This species, now renamed *Pallavicinia lyellii*, Ribbonwort or Veilwort, was not known from any other site in Scotland at that time, and is still a very rare species in Britain. Unfortunately, its current status on Lochar Moss is unknown.

Despite limited scientific equipment and ecological knowledge, the Victorians took more interest in lower plants and invertebrates than many modern naturalists. They undoubtedly knew more about the mosses at Lochar Moss than we do today, though we no longer have records of everything they found. More recently, the autumn meeting of the British Bryological (moss and liverwort) Society was held in Dumfries from 2-9 September 1961, just a few years before afforestation began. A contingent from Northumbria, with other members, visited parts of Lochar Moss where they recorded nine different species of *Sphagnum*, as well as numerous other mosses and liverworts.

In 1960, Derek Ratcliffe, (later Chief Scientist of the Nature Conservancy Council), discovered the Lochar Moss's rarest plant in the bog hollows in Racks Moss. Baltic Bog-moss *Sphagnum balticum* has only ever been recorded on seven sites in Britain, and is currently known from only three - one in Northumbria, one in Aberdeenshire, and one in Invernesshire where it was discovered as recently as 1997. Notoriously difficult to identify in the field, it requires very wet conditions. Re-found at Racks in 1962, a 1995 search reported how extensive conifer plantation meant little hope of Baltic Bog-moss survival.

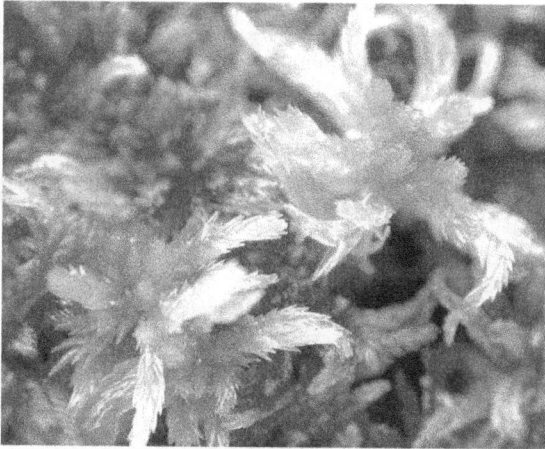

Figure 16. Baltic Bog-moss © Andy Amphlett

Flowering plants

In 1896, George Francis Scott-Elliot published '*The Flora of Dumfriesshire including part of the Stewartry of Kirkcudbright*' in which he listed species he found on Lochar Moss. Many are peatland specialists: Round-leaved Sundew, Oblong-leaved Sundew, Great Sundew, Cranberry, Bog Rosemary, Common Bladderwort and Bog Asphodel. Others are plants of wetlands and probably came from lag fen habitats on the Moss edge: Greater Spearwort, Marsh Yellow-cress, Great Yellow-cress, Marsh Cinquefoil, Purple Loosestrife, Mare's Tail, Marsh Pennywort, Tubular Water Dropwort, Nodding Bur-marigold, Trifid Bur-marigold, Scull-cap, Small Water-pepper, Unbranched

Bur-reed, Reddish Pondweed, Few-flowered Spike-rush, Star Sedge, Common Sedge, Tufted Hair-grass and Stag's-horn Clubmoss. Few plants typical of disturbed ground were mentioned in the list, but Rosebay Willowherb was *"rare"* and Small Nettle only in *"Firs near Racks"*. Either the bog vegetation was mostly in good condition, or he avoided those parts where it was damaged.

One plant on Scott-Elliot's list, Royal Fern, was suffering a double blow around the time of his writing. Not only was its habitat being drained and cultivated, but it was subject to obsessive collection in the fern-fever sweeping Victorian Britain. James M'Andrew, a Wigtownshire botanist, wrote in 1888 that *"a person told me that in her youth it was cut and dried to cover potatoes, &c., as brackens are commonly used, but that it had been carried off in cartloads by fern vendors."* The inaccessibility of Lochar Moss probably saved it from collection and clumps were still found there in the late twentieth century.

By 1940, Oleg Pullen reported on changes to the vegetation, at least on Racks Moss, due to a large colony of Black-headed Gulls. In the centre of the bog, where it was wettest, cotton-grass was being killed around the nests due to droppings, and bare, muddy ground created by gull trampling was invaded by weeds of cultivated land. In contrast to Scott-Elliot, Pullen noted Rosebay Willowherb to be "very common" here. However, immediately outside the gullery, the cotton-grass was bright green due to the added nitrogen from the droppings. Sundews survive on Lochar Moss to the present day, though only Round-leaved is easy to find. They survive the nutrient-poor bog conditions by supplementing their diet with insects trapped by means of sticky hairs on their leaves. These secrete chemicals to break down and digest the bodies. Though Round-leaved Sundew is not uncommon in Britain, the other two species are rarely encountered and so it is exceptional that the Moss supports all three. Common Bladderwort is similarly carnivorous, living as rootless, inter-twined stems suspended in water, with small bladders triggered to open by passing aquatic animals. The in-rush of water draws the animals inside. Bog

Rosemary has a strange geographical distribution. It is not uncommon on many bogs in Dumfries and Galloway, including several of the Lochar Mosses, and also occurs on bogs in Wales, north western England and, sparingly, central Scotland. But it is virtually absent from southern and eastern England and northern Scotland, even where suitable habitat appears to be present. It superficially resembles Rosemary, the fragrant herb, but the two are un-related.

Figure 17. Round-leaved Sundew © Peter Norman

Unlike most sedges which are inconspicuous and difficult to identify, Greater Tussock Sedge grows in groups of waist-high tussocks that are difficult to miss, even in the middle of winter. It is not a typical bog plant, being more characteristic of wetlands on the edge of raised bog domes. Not listed in historical records, it grows in at least two places today; between Longbridge Muir and Holmhead Moss, and on the much-modified bog at Downs Moss.

Invertebrates

In the nineteenth century, the study of insects in the Dumfries area was dominated by one man, William Lennon (1817-1899). He joined the staff of the Crichton Royal Institution, located on the edge of Lochar Moss, in 1843 and served as the personal attendant to Sir Edward Vavasour until 1885. He published extensively in the Transactions of the Dumfries and Galloway Natural History and Antiquarian Society, with butterflies, moths, beetles and bees being his main areas of interest. Amongst his records, he describes *Melanippa hastata* (Argent and Sable Moth) as *"a very pretty insect, and is rather rare but widely distributed. I found it...in Lochar Moss, near Barnkin."* However, perhaps his most notable record is his description of *Melitaea artemis* (Marsh Fritillary butterfly) as *"common in Lochar Moss."*

This most likely frequented wet areas around the edges of raised bog domes, where the caterpillar's food-plant (Devil's Bit Scabious) was found. Both species, now probably extinct on the Moss, are currently listed as Priority Species in the UK Biodiversity Action Plan, and the Marsh Fritillary is threatened across Europe.

Figure 18. Marsh Fritillary © Paul Kirkland/Butterfly Conservation

The Moss still supports a number of specialist bog invertebrates, though few have searched for them with the dedication shown by Lennon. The caterpillars of Large Heath butterfly feed on Hare's-tail Cotton Grass. It has fared rather better than Marsh Fritillary and can still be seen flitting over the moss at Longbridge Muir from mid-June to late July. As for moths, the caterpillars of Common Heath, Clouded Buff, Four-dotted Footman, Wood Tiger, Northern Eggar, and True Lover's Knot feed on heathers and associated plants, so are also to be found in heathland habitats as well as Lochar Moss and similar bogs. However, the day-flying Beautiful Yellow Underwing has recently been noted as quite numerous at Cockpool and Holmhead, even though it is rarely seen elsewhere in Dumfries and Galloway. Manchester Treble-bar is perhaps the most specialised bog moth, the caterpillars feeding on Bilberry, Cowberry, and Cranberry. It was first discovered in Britain in the early nineteenth century on Chat Moss, near Manchester. Lennon noted it, not long after, on the Lochar Moss in the 1860s and commented that *"The Manchester collectors were so elated with their success that they named it the Manchester Treble Bar. I don't see why we should not name it the Dumfries Treble Bar,* *seeing that we have it in our own locality, namely, at Tinwald Downs, where I myself found it in July last."*

Figure 19. *Helophorus tuberculatus*

From Lennon's beetle collection he notes that *Hydroporus rufifrons* (Oxbow Diving Beetle) *"is very rare; found once…in Lochar, near Sandyknowe."* Elsewhere in the UK, this beetle is currently known from less than ten sites; small, still water-bodies, fens and bog pools where they may be the dominant large invertebrate predator. Another beetle, a type of scavenger water beetle *Helophorus tuberculatus*, was recorded sometime in the nineteenth century from a fir wood near Dumfries. The location is probably Carnsalloch Moss, next to Sandyknowe, and the collector is again probably Lennon. This is now a very rare species, listed in the Red Data Book. Unfortunately, there have been no records of it since this one.

Lennon and others were already aware of changes taking place on the Lochar Moss in the nineteenth century. Lennon's obituary, written by fellow naturalist Robert Service, describes how Lennon *"used to dredge the rarest water beetles from mossy hags in Lochar where now the plough and reaper work in their seasons."*

Birds

After his medical training Sir Robert Sibbald a doctor from Fife, turned to the study of the antiquities, natural history, and topography of Scotland. He became natural historian, geographer, and physician to King Charles II, who commanded him to prepare a description of the counties of Scotland, leading to the publication of '*Scotia Illustrata, sive Prodromus Historiae Naturalis, &c.*' in 1684. Sibbald relied heavily on the maps of Timothy Pont, but to assist him in matters relating to Dumfries, he recruited the services of local medical practitioner Dr George Archibald who provided an account entitled '*Curiosities at Dumfries*'. In this, Archibald referred to "*Bittour making a great sound in the summer evenings and mornings by thrusting her beak into the ground when she cries.*" This is a reference to the 'boom' of the Bittern, a thickset, brown-streaked, secretive heron, otherwise known as Bog-drum, Miredrum, Bog-bull, Bog-hen, or Bog-trotter. Though the weird call of the Bittern is made without any thrusting of the beak into the ground, the deep resonating call can be heard from up to 3 miles away. It is perhaps understandable that residents of Dumfries perceived the muffled sound emanating from Lochar Moss as produced in the way described. Archibald also referred to 'Myresnipes', noting how "*In pleasant summer evenings they soar high in the air with a quivering voice, and are excellent meat.*" He is describing the 'drumming' of the Snipe, a whirring noise created by the vibration of the outer tail feathers during courtship display flights. They would have been resident in large numbers on the seventeenth-century Lochar Moss.

For a few of the Lochar parishes, the 'Old' and 'New' Statistical Accounts, covering the periods 1791-99 and 1834-45, provide an outline of the birdlife of that time. In the New account for Mouswald parish the "*moor buzzard* (Circus aeruginosus) *is occasionally seen in the Lochar Moss. The black-cock* (Tetrao tetrix) *and grouse* (Lagopus scoticus) *are frequently found in the moss...The wild swan* (Cygnus ferus) *and the wild goose* (Anser ferus) *is frequently shot in the Lochar Moss.*" These are references to birds now known as Marsh Harrier, Black and Red Grouse, Whooper Swan, and Bean Goose. Of these, only Whooper Swan is a regular visitor to the Lochar Moss area today. Other birds are listed in the account without location but most likely from Lochar Moss. These include Curlew, Lapwing, Golden Plover, and Snipe, all described as abundant. The Bittern was similarly described, supporting Archibald's account written over 150 years earlier. However, given this species was extinct as a breeding bird in Britain between 1886 and 1911, such a claim of abundance needs to be treated with caution. Frequently the local minister was also the best local naturalist, but perhaps the minister of Mouswald had skills of a different kind? An examination of the account from the Old Statistical Account finds a virtually identical list, except that the original writer was then much less confident about the abundance of Black Grouse and Bitterns! Perhaps, the description should be applied to the parish at the end of the eighteenth century, rather than the middle of the nineteenth.

In 1845, in Ruthwell parish, the only reference to birds on the moss was confirmation that *"some grouse are to be met with on the Lochar Moss"*. The ministers of Caerlaverock, Tinwald and Torthorwald made virtually no mention of birds but in the 1880s Chrystie published *'An Annotated List of the Birds that Breed in the Parish of Dumfries'*, listing on Lochar Moss Red Grouse and Meadow Pipit "very common"; Common Sandpiper, Curlew, Black Grouse, Grey Partridge, Cuckoo, and Reed Bunting "common"; Dunlin and Snipe "not common"; and Teal "rare". Dickie's 1898 Dumfries guidebook confirmed that Lochar Moss carried "a considerable stock of game" but did not state the species.

The first (and so far only) comprehensive description of birdlife in Dumfriesshire comes from Hugh Gladstone's 1910 *'Birds of Dumfriesshire'*. He described the *"peat-mosses"* as forming *"one of the most remarkable features of the county"* with many mentions of Lochar Moss throughout his book. The introduction describes the moss in the early nineteenth century, before being agriculturally improved, as an attractive resort for wildfowl. He quotes a report from the *Dumfries Courier* that during the extraordinarily hard winter of 1822-23 *"thousands of waterfowl found asylum there."*

Gladstone seems not to have been a regular visitor to the Moss, usually relying on reported observations of others. In a wide-ranging publication, he does not give a full species list for the site, probably omitting common birds widespread throughout the county. However, he specifically mentioned the species in Table 1.

Table 1	
Species	**Gladstone's comments relating to Lochar Moss**
Short-eared Owl	Occasionally nested amongst the tall heather *c.*1850s.
Marsh Harrier	Gladstone refers only to the Mouswald Statistical Accounts (see above).
Bittern	Formerly not-uncommon, now an accidental winter visitor.
Bean Goose	Frequently shot *c.*1832.
Whooper Swan	Coming to Lochar in 1792 and flock of 11 in February 1823, one of which was shot.
Mallard	Great numbers in severe winters.
Teal	Especially abundant.
Garganey	Several shot 1860-1885.
Goosander	Among the rarer wildfowl 1838.
Black Grouse	Formerly in considerable numbers but now rarely, if ever, seen.
Quail	None in *c.*1842, but 2-3 pairs yearly since then, increasing in recent years.
Water Rail	Fairly frequently seen.
Dunlin	A breeding species in 1881.
Black-headed Gull	800 pairs nesting at Longbridge Muir in 1908, but none by 1914. Up to 200 nests on Racks/Ironhirst, rising to 600 nests in 1914, reduced to 50 by 1921.
Lesser Black-backed Gull	200-300 nesting.
Grasshopper Warbler	Present
Reed Bunting	Present

In addition, Gladstone mentions a Shag shot about 1840, a Little Egret killed by Colonel Grierson on 2nd March 1836, a possible flock of 15 Snow Geese in March 1879, and a small number of Pallas's Sandgrouse, often seen by William Jardine during June and July 1888. However, it is the resident birds of all these accounts that are of particular note, rather than the occasional wayward seabird or migrant. Of these, the Bittern and Black

Grouse stand out. The Bittern suffered everywhere from the drainage of marshes, bogs, fens and lakes so that by 1830 it was mainly confined to East Anglia, and extinct as a British breeding bird by 1886. Though it returned in 1911, by the late twentieth century it again teetered on the edge of extinction. The fact that it survived at Lochar Moss possibly up to the end of the eighteenth century identifies this site as one of its last known refuges in Scotland. As for Black Grouse, another bird currently of national conservation concern, the accounts suggest a population present at Lochar Moss into living memory. Though Gladstone only refers to *"formerly"* considerable numbers, in one copy of *'Birds of Dumfriesshire'* there are hand-written notes in the margin (initialled, but author unknown) relating to about 20 seen there in September 1918 and several killed in November 1930.

With regard to gulls, Pullen noted an increase to about 3,000 nesting Black-headed Gulls in 1938, *"constantly persecuted by the people of Dumfries, who gather their eggs for cooking purposes."*

In the second half of the twentieth century, birds typical of peatlands continued to decline as trees began to dominate. The Moss was still sufficiently open and wet in late December 1969 that up to 2,000 Mallard were seen. Hen Harriers, which require large open habitats in which to hunt their prey of small mammals and small birds, initially benefitted from tree-planting, which dramatically increased the population of voles, and for a while they used the young plantations for communal night-time roosts. However, once the tree canopy formed, it became impossible for the harriers to hunt and they are seen only occasionally today. A few of the birds listed by Gladstone and his predecessors hang on in small numbers, including Teal, Water Rail, and Cuckoo. Recent opening up of the plantations, both for bog restoration and harvesting purposes has benefitted some of these species.

Fish

In '*A Topographical Dictionary of Scotland*', Lewis stated that the Lochar Water "*flows in a gently winding course southward, through the centre of Lochar Moss, and, deviating towards the east, falls into the Solway Frith. This river, from the level nature of the ground, has scarcely any perceptible current; it abounds with pike, perch, trout, and eels.*"

Reptiles

As well as reference to certain birds, Dr Archibald includes a section on 'Serpents' in his *c.*1684 account of '*Curiosities at Dumfries*'. This includes a Slow Worm "*at the castle meadow at Cockpool*" and "*near this and within 20 miles was a young gentleman who in a hot summer's day did readily slip into a moss and catch an adder, which he presently thrust into his bosom, and kept it there for a while without any hurt and then dismissed it.*" Whether the story is true or not (and this is not a recommended practice), the location is almost certainly Lochar Moss. Later references to Adders on Lochar Moss are frequent. Robert Service (1906), noted the following:

"*Although far from being at all scarce yet, there is no doubt that the Adder has greatly decreased from its former abundance. Drainage and reclamation have destroyed many of its haunts. The late Mr Thomas Wilkin informed me that, when a large portion of his farm of Tinwald Downs was reclaimed from Lochar Moss, a note of the Adders killed during the progress of the work was regularly kept. The average was 40 per acre.*"

Mammals

Raised bogs are not known for populations of large mammals which are more usually found in woodland. However, such animals must have wandered across Lochar Moss from time to time and peat preserves their remains well. The Red Deer antler found at the Lochar mouth and fashioned into a medieval tool may not have been from an animal living on the Moss, but another pair of antlers found on the bed of Lochar Water in the nineteenth century is likely to have originated from local wild animals, though possibly of pre-historic (pre-bog) origin. Remains of an even larger animal, an Auroch or wild ox, a species globally extinct for around 300 years, were found near Tinwald. The skull is in Dumfries Museum. The *Transactions of the Antiquarian and Natural History Society* report a clavicle and other bones from Wild Boar, found at a depth of 4.5m (15 feet) on 4th December 1862. In 1898, another was reported by W. Dickie in his Dumfries guidebook '*Dumfries and Round About*' in which he described the *"recent"* discovery of *"a portion of the hide and the shoulders of a wild boar, together with some of the bones, discovered at a depth of about 2.7m (9 feet) on the farm of Barnkin of Craigs."*

Nature Conservation

Prior to the twentieth century, the idea of conserving wildlife for intrinsic value was a completely alien concept. Even Victorian naturalists, who made detailed studies of species such as invertebrates and mosses, many of which are poorly researched today, would not have given much thought to their conservation. They noted declines of particular species, even lamented at their loss, but to campaign to protect them and their habitats from advances in farming, industrial and urban development would have been to stand in the path of progress. Scott-Elliot, who had listed the plants of Lochar Moss in 1896, was in 1919 giving speeches and writing letters encouraging its drainage. By 1953, little had changed. Evelyn Baxter and Leonora Rintoul in '*Birds of Scotland*' noted that *"the Lochar Moss, in Dumfriesshire, is a well-known resort of wild fowl"* and commented how most of the other *"old mosses"* in Scotland had been reclaimed and their fauna lost. However, at the same time as documenting such losses, Baxter and Rintoul noted the creation of *"prosperous farm land"* and *"some of the finest agricultural land"*. Their only critical comments were reserved for the direct killing of birds in the name of protection of farm stock and game birds. In the mid-1950s, the Scottish Committee of the Nature Conservancy discussed the potential of Racks

Moss as a nature reserve, but such a proposal could never seriously be considered when reclamation of such an area might offer so many other benefits.

When wildlife habitats did eventually become recognised as worthy of conservation, bogs were last in the queue for recognition. In 1964, peatland remained a habitat that was best exploited, even in the eyes of two leading Scottish conservationists. In the New Naturalist volume '*Natural History in the Highlands and Islands*', Frank Fraser Darling and John Morton Boyd described how "*sour bog plants*" occupy the ground to the exclusion of all others. They went on to describe how best to cut and dry certain kinds of peat, how to fertilise and re-seed as part of reclamation schemes, and how they were encouraged by advances in the "*development of peat lands for agriculture and forestry.*"

By the late 1970s, opinions were beginning to change. In 1977, Derek Ratcliffe, Chief Scientist of the Nature Conservancy Council, who had explored Lochar Moss and personally discovered some of its unusual flora and fauna, prepared '*A Nature Conservation Review*'. This listed the most important sites for wildlife in Britain, most of which were later selected for protection as 'Sites of Special Scientific Interest' (SSSI) under the 1981 Wildlife and Countryside Act. It included 107 peatlands with 37 in Scotland. Perhaps because much of it had already been planted with conifers or subjected to regular burning and its nature conservation value considered lost, Lochar Moss was not included in the list.

The 1980s was a decade of controversy for peatlands. Lochar Moss did not make the headlines, but afforestation of the vast blanket bogs of the Flow Country in Caithness and Sutherland did. This ignited a fierce fight between conservationists, led by Nature Conservancy Council (NCC) and RSPB, against private forestry companies. Local opinion sided with the foresters, largely because the conservation organisations were based in southern England and were seen to be interfering in northern affairs, but the balance of national public opinion, influenced by a number of effective media campaigns, supported conservation. Perhaps the deciding factor was that, unlike at Lochar Moss, no one pretended that the forests of the Flow Country were genuine attempts to make unproductive land into viable economic assets. Instead, they were clearly designed to exploit loopholes in the tax system for a small number of companies and individuals, all of whom were already very wealthy. So, on 15th March 1988, the battle came to an abrupt end when Chancellor Nigel Lawson announced closure of the loopholes in his budget speech. There was a cost to the conservationists including the break-up of the NCC, which was seen to be too powerful. But the end of new planting in the Flow Country enabled techniques of peatland restoration to begin on land previously afforested.

On 8th October 1991, the only significant area of unplanted bog on Lochar Moss at Longbridge Muir was eventually designated as an SSSI. With remarkable foresight, the designation also included the adjacent conifer plantations of Black Grain and Cockpool Moss, which had been planted in 1967-8 and 1979 respectively. In 1994, The European Union's Directive on the Conservation of Natural Habitats and Wild Flora and Fauna came into being, requiring the designation of sites of importance for particular named habitats and species of European significance to be identified as Special Areas of Conservation (SAC). Longbridge Muir qualified as such a habitat on the basis that it was an 'active raised bog', but the Directive also required inclusion of *"degraded raised bog still capable of natural regeneration"* enabling SSSI areas of Black Grain and Cockpool plantations to be added to the SAC. On 17th March 2005, these sites, together with Kirkconnell Flow National Nature Reserve on the other side of the River Nith, became the rather cumbersomely-named Solway Mosses North SAC. The remainder of Lochar Moss remains without any formal recognition of its nature conservation importance.

Restoration: "the Lochar Mosses would then represent one of the finest raised mire complexes in Western Europe."

The description above of a fully restored Lochar Moss is from a report by Richard Lindsay and Jamie Freeman of the Peatland Conservation Unit of the University of East London in 2006. These renowned peatland experts were commissioned to examine the feasibility of restoration of Racks Moss, Ironhirst Moss, and Longbridge Muir and concluded that, with the right management, an ecosystem with *"thriving mire vegetation"*, could be achieved within 30 years.

On part of Longbridge Muir, the Black Grain forestry plantation, such restoration has already been carried out. The Restoration of Scottish Raised Bogs Project 2001-2003 was led by the Scottish Wildlife Trust, supported by Scottish Natural Heritage, Forestry Commission Scotland, and others. It carried out work on 3,700 hectares of raised bogs on 11 Special Areas of Conservation costing around £1.3 million, mostly acquired through the European Union LIFE Nature Programme. As Black Grain Plantation had been designated a Site of Special Scientific Interest in 1991 and a candidate SAC, it was eligible for funding and the biggest single element of the entire national project.

In order to allow access to fell and remove the trees, floating roads had to be constructed. These needed to support the weight of the large modern forest machines as they traversed back and forwards across the Moss, but without breaking through the fragile skin of the bog into the wet, deep peat below. Trunks of the first trees to be felled formed the bed of this road, compressed tightly together and topped with 610 tons of straw. Three hundred and seventeen hectares of plantation, amounting to 45,000 cubic metres of timber, were removed and dams installed in drainage ditches to begin the process of re-wetting.

On 22nd and 23rd October 2003, Dumfries hosted a conference on bog restoration, attracting delegates from Ireland, Belgium, Germany, Ireland and Uganda, as well as throughout England,

Wales and Scotland. Many visited Longbridge Muir to look at the work of the modern forest machines and attended a conference dinner in the Cairndale Hotel. Though the purpose of the meeting was the exact opposite and the scale much reduced, similarities with the Highland and Agricultural Society's Dumfries show almost exactly 166 years previously, are obvious.

However, according to Lindsay and Freeman to restore Ironhirst, Racks and possibly Craigs Moss to the same or better condition as Longbridge Muir will not be simple. Apart from sheer cost, there are technical matters to be overcome, not all of which have been fully researched scientifically. There are also issues connected to surrounding lands. Whilst these were once part of Lochar Moss, many are drained and put to other uses by their owners.

Removal of trees from afforested areas will be required, but will not in itself result in restoration of bog vegetation and associated wildlife. Damming of the main drains to raise water levels will help, but where it has been attempted this has not been universally successful in encouraging *Sphagnum* growth on the bogs. Strangely, one of the factors that Lochar Moss has in its favour is that the tangled brash piles left after timber extraction, may help rather than hinder restoration. They act to nurse the establishment of *Sphagnum*. On other bogs with abandoned bare peat lacking such a nurse, *Sphagnum* has failed to colonise, even after 15 years.

However, Lindsay and Freeman observed signs of colonisation following clearance of Black Grain plantation on Longbridge Muir:

"The surface structure is so radically different from that normally associated with bog surfaces that there seems little possibility for such a small plant as Sphagnum *ever to overwhelm the tangle of structures...however, closer inspection reveals a pattern of small-scale structures within which* Sphagnum *and other bog species are not merely established, but are thriving and clearly expanding rapidly."*

They also advise that the surface of the peat is left with undulations no more than 40-50cm high, not dissimilar to the hummocks and hollows that would form the natural surface of the bog. This might be achieved by simply driving a tracked vehicle over the brash. Even on the extraction routes, created by densely packing very large brash to support the weight of the forestry machines during extraction, *Sphagnum* was observed growing through it after just four years.

Restoration of the former Black Grain Plantation encountered problems of birch and conifer saplings spreading across the area. In effect it may be a race to see which becomes established first, scrub cover or bog vegetation. However, the surface of Black Grain Plantation appears able to support *Sphagnum*-rich communities.
But what are the reasons for restoration of other parts of Lochar Moss?

Climate Control

The extent to which peatlands such as Lochar Moss regulate our climate is difficult, if not impossible, to accurately quantify but there is no doubt that they play a significant role. Drying of peat results in oxidisation, the peat turning into water and carbon dioxide; in effect vanishing into the atmosphere. Carbon dioxide is usually identified as the most serious of a suite of gases that contribute to the 'greenhouse effect' that is warming the earth's surface, causing a range of changes to our climate differing from place to place in ways we have yet to fully understand. Organic soils, especially peat, contain by far the majority of UK carbon stocks, much greater than found in woodlands or forests. When carbon is lost from peat (as carbon dioxide) there are serious consequences for emissions of greenhouse gases. Additionally, methane (an even more potent greenhouse gas), can be released from wet organic soils.

There is no doubt that the Lochar Moss peats store huge amounts of carbon. One survey estimated the total organic carbon stock in the top one metre of deep peat in the UK to be around 1,800 tonnes of carbon per hectare. Yet such a figure substantially underestimates the total for Lochar Moss where average peat depth is considerably more than a metre. It is not necessarily the case that organic carbon stored in the second or third metre of peat will be as high as that in the top metre, but it will certainly constitute a very significant carbon stock. To put these totals into perspective estimates for the top one metre of soil under other temperate zone terrestrial or freshwater habitats range from only 0 to 400 tonnes of carbon per hectare, often with very little, if anything, stored in deeper soils.

Changes in land-use such as drainage or cultivation that disturb soil or water-table usually result in loss of soil organic carbon into the atmosphere. This is as carbon dioxide and though erosion, as dissolved carbon in rainwater drainage and runoff. Where such land uses changes have already occurred, such eighteenth- and nineteenth-century conversion of peatland to farmland in the northern part of the moss, the quantity of carbon dioxide lost cannot ever be recaptured by tree planting or other land-use measures. Only re-creation of peat would achieve this. Even if this was possible, it would require another 8,000 years of peat formation for full recovery.

Most important is to prevent loss of carbon stocks in remaining peatlands. Should these dry out and the carbon be released into the atmosphere, it would make a massive contribution to greenhouse gas emissions. Forest growth, which on sites with low organic soil matter can lead to increased soil and tree carbon stocks, has the opposite effect on peat sites. A detailed study carried out by the Forestry Commission in 2010 on greenhouse gas (GHG) emissions from forestry on peat soils recommended:

"On wet and low fertility deep peat sites, where tree growth will be poor, restocking [replanting with trees] is likely to result in a continuing negative GHG balance. Such sites should be prioritized for open habitat restoration so that the continued loss of soil organic carbon will eventually be stopped."

Flood Mitigation

The Scottish Environment Protection Agency (SEPA) has
identified those parts of the country that are potentially
vulnerable to flooding. SEPA cannot predict precisely where and
when each flood will strike, but their assessment (concentrating
on impacts on economic assets such as buildings, roads and
farmland) indicates where there may be risk of future flooding.
Interestingly, their flood risk map for the Lochar Catchment
could double as a map of the extent of the northern part of
Lochar Moss in medieval times, with areas of potential flooding
extending right up to the western suburbs of Dumfries. Despite
centuries of drainage works, it is not surprising that those areas
where peatlands formed from wetlands 7,000 years ago remain
the most likely sites for flooding today. Fortunately for the
residents of these areas, SEPA also note that the catchment has
"high catchment flood storage and attenuation capacity." In
other words, the remaining Lochar Moss peatlands are able to
soak-up large volumes of flood-water and release it very slowly.
This reduces flood-risk in the areas around them. Should such
peatlands be lost, the flood-risk would increase.

Biodiversity

The total pre-disturbance extent of lowland raised bog in Britain
was 69,700 hectares, on 1,045 sites, 807 of them in Scotland. All
have been damaged to some extent, with only 3,800 hectares
(6%) of relatively natural bog vegetation remaining on 141 sites.
The remaining area of Lochar Moss (Racks, Ironhirst, Longbridge
Muir and Craigs) at 2,682 hectares represents the fourth largest
site in Britain. However, all three larger sites are now mostly
agricultural land with very little undisturbed peatland. The
largest surviving primary raised bog dome in Britain is the
Lochar Moss complex, more than twice the size of the next
largest area of equivalent or better quality. As Lindsay and
Freeman state, in terms of its size alone, the Lochar Moss would
be of huge international importance:

"Should it be possible to re-establish an active raised mire vegetation across these raised domes, there is no question that this would stand as one of the largest surviving raised mire complexes in western Europe."

Of course, restoration is not simply about quantity, but also quality. Would a restored Moss support the range of species that it once did? The return of species such as Baltic Bog-moss, Bittern or Black Grouse is unlikely, at least in the short-term, but restoration would undoubtedly enable the spread of surviving bog plants such as *Sphagnum*, Sundew and Bog Rosemary, together with their associated specialised insect communities.

Postscript: "The Middle of Nowhere"

In 2001, listeners to John Peel's Radio 4 programme 'Home Truths' identified the only remaining unafforested part of Lochar Moss, Longbridge Muir, as Britain's most boring place. This was based on the fact that the Ordnance Survey grid square NY0569 was marked Longbridge Muir but otherwise completely devoid of all other symbols, *"apart from the suspicion of a drainage ditch at its north-easterly extremity"*. Truly, Peel pronounced on his BBC website, here was – officially – the 'Most Boring Place in Britain'. However, Independent journalist Paul Vallely decided to visit the site himself to check whether it really was so dull. Oblivious to the geological origins of the Lochar Moss, its significance in local and national history, or the special wildlife that inhabits the bog, Vallely simply went for a walk. He noted:

"Yellow-capped mushrooms were everywhere underfoot. A solitary lapwing kur-witted overhead. In the distance, mist swirled in the setting afternoon sun on the silhouetted Criffel Hills to the west. At first, the grass just became soggier, but soon it gave way to the bog proper, a vast expanse of sheer heather in big brown tufts between the glinting pools of peat-dark water. On and on it stretched, with only the odd stunted birch tree to relieve the vastness, and a skylark's hovering song to punctuate the keen listening silence. Finally, after an hour's

painstaking walk, it was as if a physical journey became a metaphysical one. It was one of those places where heaven and earth touch. Boring, it transpires, is in the eye of the beholder."

References

Aiton, W. (1805) *A Treatise on the Origin, Qualities and Cultivation of Moss-earth.* Glasgow.

Anon. (1866) Report of Field Meeting. Tree Found in Lochar Moss, Syke, Torthorwald. *Transactions of Dumfriesshire & Galloway Natural History & Antiquarian Society, Series I,* **Vol. 2,** 13-18.

Anon. (1881) List of Specimens, Books, &c., belonging to the Society. *Transactions of Dumfriesshire & Galloway Natural History & Antiquarian Society, Series II,* **Vol. 2,** 85.

Burnett, T.R. (1949) Notes on some human bones found in Lochar Moss. *Transactions of Dumfriesshire & Galloway Natural History & Antiquarian Society, Series III,* **Vol. 26,** 126-7.

Bragg, O.M., Lindsay, R.A., Robertson, H., *et al.* (1984*) An Historical Survey of Lowland Raised Mires, Great Britain. Report to the Nature Conservancy Council.* Nature Conservancy Council, Peterborough.

Brooks, S., & Stoneman, R. (1997) *Conserving Bogs: the Management Handbook.* The Stationery Office, Edinburgh.

Cowie, T., Picken, J., & Wallace, C. (2011) Bog Bodies from Scotland: Old Finds, New Records. *Journal of Wetland Archaeology,* **10,** 1-45.
Cruickshank, J. (1842) List of Jungermannia &c. observed in the neighbourhood of Dumfries.*The Phytologist* (1843).

Department of Agriculture and Fisheries for Scotland (1964) *Scottish Peat Surveys, Volume 1, South West Scotland.* HMSO, Edinburgh.

Dickie, W. (1898) *Dumfries and Round About*. Third edition. Dumfries.

ECOSSE (2007) *Estimating carbon in organic soils sequestration and emissions*. Scottish Executive, Edinburgh.

The Farmers Magazine, 7th, July-December 1837, p122.

Fenton, A. (1976) *Scottish Country Life*. John Donald, Edinburgh.

Fraser Darling, F., & Morton Boyd, J. (1964) *The Highlands and Islands*. Collins, London.

Gladstone, H.S. (1910) *Birds of Dumfriesshire*. Witherby & Co., London.

Grahame, J. (1809) *British Georgics*. J. Ballantyne & Co., Edinburgh.

Gray, P. (1842) A list of the rarer Flowering Plants and Ferns of the neighbourhood of Dumfries, with remarks on the physical conditions of the district. *The Phytologist*: Second Annual Part (1843), London, 416-419.

Groome, F.H. (1882-1885) *Ordnance Gazetteer of Scotland: A Survey of Scottish Topography, Statistical, Biographical and Historical*. Thomas C. Jack, Grange Publishing Works, Edinburgh.

Hansard (1949) Vol.162 cc1201-2WA 1202WA.

Hume, D. (ed.) (1834) *Decisions of the Court of Session 1781-1822*. William Blackwood & Sons, Edinburgh and Thomas Cadell, London.

Irish Engineers Journal Supplement (1970), p23-27.

Jardine, W.G. (1975) Chronology of Holocene marine transgression and regression in south-west Scotland. *Boreas*, 4.

Jardine, W.G., & Masters, L.J. (1977) A dug-out canoe from Catherinefield Farm, Locharbriggs, Dumfriesshire. *Transactions of Dumfriesshire & Galloway Natural History & Antiquarian Society, Series III,* **Vol. 52**, 56-65.

Johnston, B. (1794) *General View of the Agriculture of Dumfries.* London.

Kerr, W.A. (1905) *Peat and Its Products: An Illustrated Treatise on Peat and Its Products as a National Source of Wealth.* Begg, Kennedy & Elder, Glasgow.

Lennon, W. (1865) Notes on a few of the Rare Lepidoptera observed in the Vicinity of Dumfries. *Transactions of Dumfriesshire & Galloway Natural History & Antiquarian Society, Series I,* **Vol. 2**, 62-65.

Lennon, W. (1881) Notes on Rare Beetles. *Transactions of Dumfriesshire & Galloway Natural History & Antiquarian Society,* Series II, **Vol. 2**, 77.

Lewis, S. (1846) *A Topographical Dictionary of Scotland.* S. Lewis and Co, London.

Lindsay, R.A. & Freeman, J. (2006) The Lochar Mosses: Present Condition and Future Potential. University of East London, London.

Lindsay, R. (2010) Peatbogs and Carbon: A Critical Synthesis. Environmental Research Group, University of East London, London.

Macdonald, J. (1906) Dr Archibald's "Account of the curiosities of Dumfries" and "Account about Galloway". *Transactions of Dumfriesshire & Galloway Natural History & Antiquarian Society, Series II,* **Vol. 17**, 50-63.

Maclean, M. (ed.). (1901) *Archaeology, Education, Medical, & Charitable Institutions of Glasgow*. Local Committee for the Meeting of the British Association for the Advancement of Science, Glasgow.

Marshall, J.R. (1961) Investigations into some aspects of the physiography of the upper Solway marshes and mosses. Unpublished NCC files quoted in Bragg, O.M., Lindsay, R.A., Robertson, H. *et al.* (1984) *An Historical Survey of Lowland Raised Mires, Great Britain*. Report to the Nature Conservancy Council. Nature Conservancy Council, Peterborough.

Maxwell, R. (ed.) (1743) *Select Transactions of The Honourable The Society of Improvers in the knowledge of Agriculture in Scotland*. Edinburgh.

McBride, A. (2003) *Grazing on Lowland Raised Bogs. Summary of talk delivered at Restoration of Scottish Raised Bogs Conference*, Dumfries 22-23 October 2003. Scottish Wildlife Trust, Edinburgh.

McClumpha, I. (2011) *A Look at the Locharwoods*. Self-publishing 35-page booklet, Lockerbie, Dumfriesshire, Scotland,

McDowall (1867) *History of the Burgh of Dumfries*. Adam & Charles Black, Edinburgh.

M'Andrew, J. (1890) Notes on the Flora of Wigtownshire. *Transactions of the Dumfries and Galloway Natural History and Antiquarian Society, Series II*, **Vol. 6**, 17.

M'Diarmid, W. (1866) On the Recent Discovery of a Stone Coffin in Lochar Moss near Tinwald Downs. *Transactions of Dumfriesshire & Galloway Natural History & Antiquarian Society, Series I*, **Vol. 2**, 2.

Morison, J., Vanguelova, E., Broadmeadow, S., Perks, M., Yamulki, S., & Randle, T. (2010) *Understanding the Greenhouse Gas implications of forestry on peat soils in Scotland*. Forestry

Commission Research Report, Forestry Commission Scotland, Edinburgh.

Morison, J., Matthews, R., Miller, G., Perks, M., Randle, T., Vanguelova, E., White, M., & Yamulki, S. (2012) *Understanding the carbon and greenhouse gas balance of forests in Britain*. Forestry Commission Research Report, Forestry Commission, Edinburgh.

Pagan, W. (1865) *The Birthplace and Parentage of William Paterson*. Ballantyne & Co., Edinburgh.

Pickin, J. (2004) Bog bodies from Dumfries and Galloway, *Transactions of Dumfriesshire & Galloway Natural History & Antiquarian Society, Series III*, **Vol. 78**, 31-36.

Scott-Elliott, G.F. (1896) *The Flora of Dumfriesshire including part of the Stewartry of Kirkcudbright*. J. Maxwell & Son, Dumfries.

Scott-Elliott, G.F. (1921) The Plants of Holms, Merselands, and River Valleys. *Transactions of Dumfriesshire & Galloway Natural History & Antiquarian Society, Series III*, **Vol. 7**, 32-53.

Service, R. (1906) The Pre-Historic Red Deer of Solway. *Transactions of Dumfriesshire & Galloway Natural History & Antiquarian Society, Series II*, **Vol. 17**, 309.

Singer, W. (1812) *General View of the Agriculture, State of Property and Improvements in the County of Dumfries*. Ballantyne and Co, Edinburgh.

Smout, T.C. (2009) *Bogs and People in Scotland since 1600*. In: *Exploring Environmental History*. Selected Essays. Edinburgh University Press, Edinburgh.

Steele, A. (1826) *The Natural and Agricultural History of Peat-moss or Turf-bog*. W. & D. Laing & Adam Black, Edinburgh.

Wallace, R. (1918) The Lower Nith in its relation to Flooding and Navigation. *Transactions of Dumfriesshire & Galloway Natural History & Antiquarian Society, Series III,* **Vol. 5**, 128-136.

Williams, J. (1966) A Sample of Bog Butter from Lochar Moss, Dumfriesshire. *Transactions of Dumfriesshire & Galloway Natural History & Antiquarian Society, Series III,* **Vol. 43**, 153-4.

Archival Sources

Dumfries Archive Centre GG2/7/2 (1524) Warrant by James V to Summon Arbiters to Settle Dispute over Lochar Moss, part of the Common Good of the Royal Burgh of Dumfries.

Dumfries Archive Centre RB2/2/118 (1725) Report by Provost Edgar of Dumfries to the Commissioners of Supply about a Bridge over the Lochar.

Dumfries Archive Centre RB2/2/115 (1728) Contract with William Hanna and William Wood to Build a Bridge over the Lochar.

Dumfries Archive Centre RB2/2/119 (1728) Payment to William Hanna and William Wood for Building a Bridge over the Lochar.

Dumfries Archive Centre (1762) Mr J. Smeaton's Observations &c anent draining Lochar Moss and Estimate thereof.

Dumfries Archive Centre GGD40/2 (1795) Rental of the Baronies of Mouswald, Torthorwald, Tinwald and Craigs and the Feus of Carzield.

Dumfries Archive Centre GGD498/4/8 (1799) An account by the Town Council to the Town Clerks of Dumfries, Simon Mackenzie and Mr Maxwell for coals and peat.

Dumfries Archive Centre GG2/7/16A (1825) Sasine to Dumfries of its Common Good Lands from Trustees Holding them against the Burgh's Debts.

Dumfries Archive Centre GGD37/10/7 (1845) Regulations to be observed by all who cast peats on Craig Moss.

Dumfries Archive Centre GGD131/B2/61 (mid 1800s) Walter Newall Architect: Plan of the Geology of the Solway and Lochar Moss.

Dumfries Archive Centre GGD131/F5/3 (mid 1800s) Walter Newall's Maps of the Estates along the Lochar River, Dumfriesshire.

Dumfries Archive Centre (1827) State of facts relative to the Water of Lochar and Moss thereof.

Dumfries Archive Centre MP75 (1878) Plans for Improvement to River Lochar.

Dumfries Archive Centre (1919-1921) Sir James Crichton-Browne's Letters and Notes relating to Lochar Drainage.

Dumfries Archive Centre FEconDev IN13/2 (1980-1996) Peat Development Project file of the Department of Economic Development of the Dumfries and Galloway Regional Council

Dumfries Archive Centre FEconDev IN13/1 (1980-1996) Peat Extraction Project file of the Department of Economic Development of the Dumfries and Galloway Regional Council

The Royal Society JS/6/128 (*c.*1754) John Smeaton's Plan of Locker Moss, Dumfriesshire

Notes

Remembering the Solway

Chris Spencer

The Solway Wetlands Landscape Partnership

The project 'Remembering the Solway' was set up in 2013 as one of twenty-nine initiatives within the Solway Wetlands Landscape Partnership, a Heritage Lottery Fund supported scheme to restore, protect and celebrate the landscape of the Solway Plain in North Cumbria. A key aspiration of the scheme was to capture and record some of the memories of people living on the Solway throughout the last century, in order to find out what life was like in this area within living memory, and to what extent it has changed over time.

The project started in 2015, when a small group of interested local people came together, keen to develop this project and felt it was important to collect these memories before they were eventually lost. Once the group were trained in the practice of oral history, they set about identifying people to talk to and set about recording the interviews. The group met fortnightly in the Port Carlisle Chapel.

The area that the group decided to focus on was the central, northern section of the Solway Plain, encompassing the villages of Newton Arlosh, Kirkbride, Bowness-on-Solway, Port Carlisle, Drumburgh and Burgh-by-Sands. This contains the largest concentration of lowland raised mire (Moss) and is an area of low-lying coastal farmland interspersed with small villages and hamlets. In total, 46 interviews were carried out, recording the memories of 53 people.

The conference presentation is by volunteers from the Remembering the Solway Oral History Group and Naomi Hewitt, Assistant Manager of the Solway Coast AONB and former member of the Solway Wetlands Landscape Partnership Scheme.

More information on the Solway wetlands Landscape Partnership including the Remembering the Solway book and film can be found at www.solwaywetlands.org.uk

The 2017 conference Field Trip

Our conference field visit took us to Bowness Common. This is an extensive lowland raised mire, part of the South Solway Mosses National Nature Reserve, and which incorporates part of RSPB Campfield Marsh Nature Reserve.

Our 5-km walk took us up to the viewpoint at the Rogersceugh Drumlin situated in the centre of the bog (there was a short side excursion on the way for the more adventurous to view the restoration works along the line of the railway). This was a very interesting but very wet site. From here the line of the railway was visible along with the topography of the bogs in relation to drained farmland, historical drainage and rewetting. We then went north across the bog via a network of boardwalks via historical domestic cutting sites and through the RSPB wet farmland to end up at the Solway Wetlands Centre. Joining us on the field trip were Ann Lingard ('Crossing the Moss'), Alasdair Brock (Natural England NNR Manager), and Dave Blackledge (RSPB Site Manager).

The field trip was organised by Chris Spencer, Project Manager of the Heritage Lottery-funded Solway Wetlands Landscape Partnership Scheme. Solway Wetlands has worked across the Solway area to restore, protect and celebrate this distinctive landscape, at RSPB Campfield Reserve the scheme has improved access, visitor facilities, interpretation, funded wetland restoration works and engaged the community in learning and conservation activity.

Crossing the Moss: The Story of Bowness Common and the Solway Junction Railway

Ann Lingard & James Smith

www.annlingard.com & www.jamessmithphotography.co.uk

In 1865, the Brogdens and their engineer James Brunlees started a major infrastructure project – to build the Solway Junction Railway (SJR) from England to Scotland across the Solway Firth, to transport West Cumbrian haematite quickly and cheaply to the furnaces in Lanarkshire.

The fact that the SJR would have to be constructed across the raised mire of Bowness Moss (Common) on the Cumberland side as well as over the notoriously chaotic Solway Firth didn't put them off at all. But the peat of Bowness Common fought back. When work started, "Horses could not go upon it and, except in the height of summer, cattle could not even traverse it.". Navvies laboured to cut channels to drain the peat: "water ran in river-like streams" and the level of the peat dropped by 4-5 feet each side of the track. Wooden faggots, extra-long sleepers, near-disastrous test-runs of engines ... not until the summer of 1870 was the track over Bowness Moss passed as fit for passenger as well as freight traffic.

But during the next decade or so the viaduct suffered mishaps and disaster, the railway proved uneconomic, and was eventually closed in 1926. In 1934, the track across the Common was dismantled. Then, at the end of the last millenium, the RSPB and Natural England started to acquire the precious Common and began the complex (and visually-dramatic) programme of re-wetting that continues to this day.

Photographer James Smith and Ann Lingard became intrigued by the 'story' of Bowness Common – a story which so few people know, of the 150 years of human intervention on the 'Moss'. It took us to archives, websites, railway 'chatrooms', to the offices of the RSPB and Natural England – and on 'expeditions' along the stub of the embankment and across the boggy dome of the Common. James squelched through bogs and sent his drone off across the Common and the Firth, hoping that it would return with its own, visual, stories (it did).

You can now read the story of Bowness Moss and the SJR, and enjoy James' stunning aerial video, in *Crossing the Moss,* www.crossingthemoss.wordpress.com.

We are very grateful to the Solway Wetlands Landscape Partnership and the HLF for their support, and to all the local volunteers who took part in the project.

Crossing the Esk and the bogs 1847 © Ian Rotherham

Summary & Conclusions

Ian D. Rotherham

This short volume provides a first overview of the history and heritage of Cumbria's peat bogs and of those in the areas around the English Lakes. Much of this is an often overlooked heritage and importantly, this insight has major implications for restoration projects. However, it is clear from this collection of chapters that there is much that still remains to be done with regard to peatland heritage generally, but Cumbrian peat bogs specifically.

Although at long last there are significant projects to restore and re-wet peatlands, there are almost no projects conserve or celebrate peatland heritage. This is itself a shame since local history has proved a valuable way to engage a much wider audience with peat issues and peat bog restoration.

Finally, projects such as 'Cumbria BogLife' and 'Solway Connections' demonstrate what can be achieved with relatively modest resources. However, the other take-home message from projects of this sort is the need for project champions and peat bog stakeholders to be supported sustainably. Furthermore, as raised at the closing event for Cumbria BogLife, we do need to see a genuine and demonstrable ending of the use of peat and peat products in gardening and in the horticultural industry.

The two-day seminar was hugely successful, great fun, and amongst other things it showed how much potential knowledge and enthusiasm are there to be tapped with regards to the history and heritage of peat, peat bogs, and their history of exploitation.

REMEMBERING THE SOLWAY

A project to collect, share and celebrate memories of life and landscape on the Solway Plain

solwayWetlands
LANDSCAPE PARTNERSHIP SCHEME

heritage
lottery fund
LOTTERY FUNDED

War & Peat

Ian D. Rotherham & Christine Handley (eds)

THE
LOST
FENS

ENGLAND'S GREATEST ECOLOGICAL DISASTER

IAN D. ROTHERHAM

Conference field visit – photographs © Solway Connections Guided
Heritage Tours

www.ingramcontent.com/pod-product-compliance
Lightning Source LLC
Chambersburg PA
CBHW061240220326
41599CB00028B/5492